Yamaha
XT, TT & SR 500
Singles
Owners
Workshop
Manual

by Mansur Darlington

with an additional Chapter on the 1979 to 1983 models
by Mark Coombs

Models covered
XT500. 499cc. UK July 1976 to April 1983
XT500. 499cc. US September 1975 to 1981
TT500. 499cc. US only September 1975 to 1981
SR500. 499cc. UK April 1978 to April 1983
SR500. 499cc. US September 1977 to 1981

ISBN **978 1 85010 749 1**

© Haynes Group Limited 1992

(342-12R8)

Haynes Group Limited
Haynes North America, Inc

www.haynes.com

British Library Cataloguing in Publication Data
Darlington. Mansur Yamaha XT, TT & SR500 singles owners workshop manual. 1. Motorcycles. Maintenance & repair I. Title II. Coombs, Mark *1969-* III. Series 629.28775 ISBN 1-85010-749-3
Library of Congress Catalog Card Number
90-85498

Acknowledgements

Our thanks are due to Mitsui Machinery Sales (UK) Limited, who gave permission to use the line drawings used throughout this manual, and to Will Wells of East Chinnock, Somerset who supplied the XT500C featured in the photographs which accompany the text.

Our thanks are also due to Jim Patch of Yeovil Motor cycle Services, who supplied us with technical information. Pete Wareham of Sparkford kindly allowed us to photograph his XT500, shown on the front cover.

Brian Horsfall gave considerable assistance with the strip-down and rebuilding and devised the ingenious methods for overcoming the lack of service tools. Les Brazier arranged and took the photographs that accompany the text. Jeff Clew edited the text.

We should also like to thank the Avon Rubber Company who kindly supplied us with information and advice about tyre fitting; NGK Spark Plugs (UK) Ltd for information and photographs relating to sparking plug conditions, and Renold Limited for advice on chain care and renewal.

About this manual

The author of this manual is convinced that the only way a meaningful and easy to follow text can be written, is to do the work himself, under conditions that exist in the ordinary household. As a result, the hands seen in the photographs are those of the author. The machine is not new, having covered a few thousand miles. Therefore the conditions encountered are the same as those found by the average owner/rider who through following the text can save himself costly repair bills. Yamaha service tools have not been used. We have proved that there are ways of removing or slackening vital components when special tools are not available, provided this is done carefully and a reasonable amount of time is allowed. Risk of damage must be avoided at all costs.

Each of the 6 Chapters is divided into numbered Sections. Within the Sections are numbered paragraphs. In consequence cross reference throughout the manual is both straightforward and logical. When a reference is made "See Section 2.8" it means Section 2, paragraph 8 in the same Chapter. If another Chapter were meant the text would read "See Chapter 6, Section 2.8".

All photographs are captioned with a Section/paragraph number to which they refer and are always relevant to the Chapter text adjacent. Figure numbers (usually line drawings) appear in numerical order, within a given Chapter. Fig. 1.1 therefore refers to the first figure in Chapter 1. Left-hand and right-hand descriptions of the parts of the machines apply when the rider is seated on the machine in the normal riding position.

Motorcycle manufacturers continually make changes to specifications and recommendations, and these, when notified, are incorporated into our manuals at the earliest opportunity.

We take great pride in the accuracy of information given in this manual, but motorcycle manufacturers make alterations and design changes during the production run of a particular motorcycle of which they do not inform us. No liability can be accepted by the authors or publishers for loss, damage or injury caused by any errors in, or omissions from, the information given.

Contents

Right-hand view of 1975 Yamaha XT500C

Right-hand view of 1978 Yamaha XT500E

Right-hand view of 1978 Yamaha SR500

Close-up of 1978 Yamaha SR500 engine

Introduction to the Yamaha XT, TT and SR500 singles

Although the history of Yamaha can be traced back to the year 1887, when a then very small company commenced manufacture of reed organs, it was not until 1954 that the company became interested in motorcycles. As can be imagined, the problems of marketing a motorcycle against a background of musical instruments manufacture were considerable. Some local racing successes and the use of hitherto unknown bright colour schemes helped achieve the desired results and in July 1955 the Yamaha Motor Company was established as a separate entity, employing a work force of less than 100 and turning out some 300 machines a month.

Competition successes continued and with the advent of tasteful styling that followed Italian trends, Yamaha became established as one of the world's leading motorcycle manufacturers. Part of this success story is the impressive list of Yamaha 'firsts' – a whole string of innovations that include electric starting, pressed steel frame, torque induction and 6 and 8 port engines. There is also the "Autolube" system of lubrication, in which the engine-driven pump is linked to the twist grip throttle, so that lubrication requirements are always in step with engine demands.

In introducing the TT, XT and SR series of 500cc machines, Yamaha have scored another 'first'. Although the concept of a large capacity single cylinder four-stroke is by no means new – this type of machine having dominated the British motorcycle market for many years – Yamaha is the first manufacturer from Japan to re-appraise the benefits of the big single, and offer an example to the public.

The TT500 model and the XT500 were introduced to cater for the competition rider, and the rider who requires both off-road and road transport. Soon after the introduction of these two models, public opinion indicated that a pure road machine would be welcome, embodying the characteristics of good low-speed torque, light weight and simplicity of construction found in so many of the British singles produced earlier. The SR500 introduced in 1978 combines some of these traits with a Japanese style and additional sophistications.

For information on 1979 to 1983 models refer to Chapter 7.

Dimensions and weights

	TT500C	TT500D and E	XT500C	XT500D and E	SR500 and SR500E
Overall length	2,110 mm (83.1 in)	2,119 mm (83.4 in)	2,170 mm (85.4 in)	2,155 mm (84.8 in)	2,105 mm (82.9 in)
Overall width	935 mm (36.8 in)	904 mm (35.6 in)	875 mm (34.4 in)	875 mm (34.4 in)	835 mm (32.9 in)
Overall height	1,120 mm (44.1 in)	1,136 mm (44.7 in)	1,220 mm (48.0 in)	1,180 mm (46.5 in)	1,150 mm (45.3 in)
Wheelbase	1,420 mm (55.9 in)	1,426 mm (56.1 in)	1,420 mm (55.9 in)	1,420 mm (55.9 in)	1,410 mm (55.5 in)
Minimum ground clearance	215 mm (8.5 in)	223 mm (9.2 in)	215 mm (8.5 in)	225 mm (8.9 in)	165 mm (6.5 in)
Dry weight	119 kg (262 lb)	123 kg (271 lb)	138 kg (304 lb)	139 kg (306 lb)	163 kg (359 lb)

Ordering spare parts

When ordering spare parts for the Yamaha XT, TT and SR500 series, it is advisable to deal direct with an official Yamaha agent, who will be able to supply many of the items required ex-stock. Although parts can be ordered from Yamaha direct, it is preferable to route the order via a local agent even if the parts are not available from stock. He is in a better position to specify exactly the parts required and to identify the relevant spare part numbers so that there is less chance of the wrong part being supplied by the manufacturer due to a vague or incomplete description.

When ordering spares, always quote the frame and engine numbers in full, together with any prefixes or suffixes in the form of letters. The frame number is found stamped on the right-hand side of the steering head, in line with the forks. The engine number is stamped on the right-hand side of the crankcase, immediately behind the cylinder barrel.

Use only parts of genuine Yamaha manufacture. A few pattern parts are available, sometimes at cheaper prices, but there is no guarantee that they will give such good service as the originals they replace. Retain any worn or broken parts until the replacements have been obtained; they are sometimes needed as a pattern to help identify the correct replacement when design changes have been made during a production run.

Some of the more expendable parts such as sparking plugs, bulbs, tyres, oils and greases etc., can be obtained from accessory shops and motor factors, who have convenient opening hours, and can often be found not far from home. It is also possible to obtain parts on a Mail Order basis from a number of specialists who advertise regularly in the motorcycle magazines.

Location of frame number

Location of engine number

Routine maintenance

Refer to Chapter 7 for information relating to the 1979 to 1983 models

Periodic routine maintenance is a continuous process that commences immediately the machine is used. It must be carried out at specified mileage recordings, or on a calendar basis if the machine is not used frequently, whichever is the sooner. Maintenance should be regarded as an insurance policy, to help keep the machine in the peak of condition and to ensure long, trouble-free service. It has the additional benefit of giving early warning of any faults that may develop and will act as a regular safety check, to the obvious advantage of both rider and machine alike.

The various maintenance tasks are described under their respective mileage and calendar headings. Accompanying diagrams are provided, where necessary. It should be remembered that the interval between the various maintenance tasks serves only as a guide. As the machine gets older or is used under particularly adverse conditions, it would be advisable to reduce the period between each check. Some of the tasks are described in detail, where they are not mentioned fully as a routine maintenance item in the text. If a specific item is mentioned but not described in detail, it will be covered fully in the appropriate Chapter. No special tools are required for the normal routine maintenance tasks. The tools supplied with every new machine will prove adequate, in most cases, but if additional tools are needed, it is advisable to purchase only those of good quality, which will not damage any parts of the machine on which they are used.

Weekly, or every 200 miles (300 kilometres)

---

1 Tyre pressures

Check the tyre pressures with a pressure gauge that is known to be accurate. Always check the pressures when the tyres are cold. If the tyres are checked after the machine has travelled a number of miles, the tyres will have become hot and consequently the pressure will have increased, possibly as much as 8 psi. A false reading will therefore always result.

The recommended tyre pressures are:

	TT500C, D and E	XT500C, D and E	SR500
Front:			
Off road	13 psi	13 psi	-
*On road	-	18 psi	26 psi
Rear:			
Off road	16 psi	16 psi	-
*On road	-	21 psi	28 psi

*When carrying a passenger or travelling at continuous high speeds increase the front tyre pressure by 2-3 psi, and the rear tyre pressure by 3-4 psi.

2 Engine oil level

Check the level of oil in the oil tank by using the combined filler cap/dipstick located just forward of the petrol tank. Ensure that the machine is upright when the check is made. When making this check the filler cap should not be screwed fully home; it should be rested on top of the threads in the filler neck. If required, replenish the oil reservoir with SAE 20W/50 engine oil. Do not run the engine with the oil level lower than the minimum level line, and do not overfill the oil tank.

Oil level should be within criss-cross area

3 Safety check

Give the machine a close, visual inspection, checking for loose nuts and fittings, frayed control cables, etc.

4 Legal check

Ensure that the lights, horn and traffic indicators function correctly, also the speedometer. If any bulbs have to be renewed, make sure they have the same rating as the original. Remember that if different wattage bulbs are used in the traffic indicators, the flashing rate will be altered.

Monthly, or every 500 miles (800 kilometres)

Complete the tasks listed under the weekly/200 mile heading and then carry out the following checks.

1 Tyre damage

Rotate each wheel and check for damage to the tyres,

especially splitting on the sidewalls. Remove any stones or other objects caught between the treads. This is particularly important on the front tyre, where rapid tyre deflation due to penetration of the inner tube will almost certainly cause total loss of control of the machine.

2   Spoke tension

Check the spokes for tension, by gently tapping each one with a metal object. A loose spoke is identifiable by the low pitch noise emitted when struck. If any one spoke needs considerable tightening, it will be necessary to remove the tyre and inner tube in order to file down the protruding spoke end. This will prevent it from chafing through the rim band and piercing the inner tube.

In the case of cast aluminium alloy wheels, as fitted to the SR500 model, they should be examined very carefully for any signs of cracking or other damage.

Cracking or splitting is most likely to occur at the point where the spokes join either the rim or the hub.

3   Front brake adjustment – drum brake models

Adjust the front brake cable so that there is about 5-8 mm (0.2-0.3 in) of free play measured between the lever stock face and the lever, before the brake starts to bite. Adjustment may be carried out using the cable adjuster at either end of the cable. It is normal practice to use the lower adjuster for initial adjustments and the adjuster on the handlebar lever for finer running adjustment.

4   Rear brake adjustment – drum brake models

When the rear brake is in correct adjustment the total brake pedal travel measured at the toe tread should be within the range 20-30 mm (0.8-1.2 in). If the travel is greater or less than this carry out the necessary adjustment by means of the shouldered nut at the brake arm end of the cable.

5   Hydraulic fluid level – disc brake models

Check the level of the hydraulic fluid in the master cylinder reservoir mounted on the handlebars. The level can be seen through the transparent reservoir and should be between the upper and lower level marks. Ensure that the handlebars are in the central position when a level reading is taken and also when the cap and diaphragm are removed. Replenish the reservoir with an hydraulic fluid of the recommended specification.

In the case of SR500E models, this will also apply to the rear brake master cylinder reservoir, which is found under the right-hand side cover, immediately above the rear brake pedal. It is imperative that the correct grade of fluid is used.

In normal use the fluid level will drop very slowly, only if a leak occurs will the reduction in level be noticeable over a short period.

Apply graphite lubricant to final drive chain

6   Final drive chain lubrication

In order that final drive chain life can be extended as much as possible, regular lubrication and adjustment is essential. This is particularly so when the chain is not enclosed or is fitted to a machine transmitting high power to the rear wheel. The chain may be lubricated whilst it is in place on the machine by the application of one of the proprietary chain greases contained in an aerosol can. Ordinary engine oil can be used, though owing to the speed with which it is flung off the rotating chain, its effective life is limited.

The most satisfactory method of chain lubrication can be made when the chain has been removed from the machine. Clean the chain in paraffin and wipe it dry. The chain can now be immersed in one of the special chain graphited greases. The grease must be heated as per the instructions on the can so that the lubricant penetrates into the areas between the link pins and the rollers.

The exact intervals at which the chain will require relubrication is largely dependent on the conditions in which the machine is used. It follows that the chain of a machine used continuously in wet and muddy off-road pursuits will require attention more often than that of a road machine used only in the dry.

7   Final drive chain adjustment

Check the slack in the final drive chain. The correct up and down movement, as measured at the mid-point of the chain lower run, should be 15-20 mm (0.6-0.8 in) on SR models, and 30-40 mm (1.2-1.6 in) on XT and TT models. Adjustment should be carried out as follows: place the machine on the centre stand so that the rear wheel is clear of the ground and free to rotate. Remove the split pin from the wheel spindle and slacken the wheel nut a few turns. On those machines with a drum rear brake, the split pin should be removed from the brake torque rod rear bolt, and the nut slackened. Loosen the locknuts on the two chain adjuster bolts. Rotation of the adjuster bolts in a clockwise direction will tighten the chain. Tighten each bolt a similar number of turns so that wheel alignment is maintained. This can be verified by checking that the mark on the outer face of each chain adjuster is aligned with the same aligning mark on each fork end. With the adjustment correct, tighten the wheel nut and fit a new split pin. Finally, retighten the adjuster bolt locknuts.

8   Battery electrolyte level – except TT models

Detach the frame left-hand side cover so that access may be gained to the battery. The electrolyte solution should be between the upper and lower level lines. If the electrolyte solution is low it should be replenished, using distilled water. This is best done when the battery is removed from the machine. Disconnect the positive lead and the negative lead, release the battery strap and breather tube, and then lift the battery from place.

Two-monthly, or every 1000 miles (1600 kilometres)

Complete each of the checks listed under the weekly/200 mile and monthly/500 mile headings, then carry out the following additional checks:

1   Air filter cleaning – SR500 models

The air cleaner box is located below the nose of the dualseat. To gain access to the box, remove the right-hand frame sidecover, which is retained by a single bolt at the lower edge and on hooks projecting from the frame at the upper edge.

Detach the side cover from the air cleaner box after removing the four retaining screws. The air filter element is a sliding fit in the box, located by a spring steel clip.

The element is cleaned by blowing through it from the inside with compressed air or by lightly tapping it so that loose dust on the surface will be displaced.

If the element is damaged or is contaminated badly in any

way, it should be renewed. A blocked filter will increase fuel consumption and a perforated one may give rise to a weak mixture, resulting in a hot running engine. The filter is marked 'top' and 'front' and should be refitted accordingly.

Do not on any account run the machine with the air filter removed or with the air cleaner hoses disconnected. If this precaution is not observed, the engine will run with a permanently weak mixture, which will cause overheating and possible seizure.

2 Control cable lubrication

Use motor oil or an all-purpose oil to lubricate the control cables. A good method for lubricating the cables is shown in the accompanying illustration, using a plasticine funnel. This method has a disadvantage in that the cables usually need removing from the machine. An hydraulic cable oiler which pressurises the lubricant, overcomes this problem. Do not lubricate nylon lined cables as the oil will cause the nylon to swell, thereby causing total cable seizure.

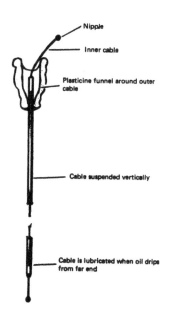

Oiling a control cable

3 Greasing points

Apply a grease gun containing a multi-purpose grease to the swinging arm grease nipple. Other points which require occasional greasing are the rear brake pedal pivot, the brake rod clevis pins, the side stand pivot and centre stand pivot and the rear brake cam.

Quarterly or every 2000 miles (3000 kilometres)

Complete all the checks in the foregoing maintenance sections and then carry out the following:

1 Changing the engine/transmission oil

Remove the sump guard (where fitted) and position a container that will hold at least 2 litres (3¼ pints) below the crankcase. Remove the drain plug from the rear of the sump cover and allow the oil to run out. To aid draining, unscrew the small bleed screw from the top of the oil filter chamber in the right-hand engine cover. After draining is complete, refit the plug, ensuring that the sealing washer is in good condition.

Move the drain pan forwards so that it is below the front of the frame downtube, and then remove the drain plug from near the base of the tube. Unscrew the filler cap to aid drainage. Be prepared for the oil to gush out and fall initially some distance from the frame tube.

After drainage is complete, refit and tighten the plug. Draining the oil should be carried out when the engine is hot; the warm oil will flow more easily and drain more completely.

Replenish the oil reservoir with approximately 2.0 litres (4.2/3.7 US/Imp pints) of SAE 20W/50 engine oil.

To check that the oil is circulating correctly and that the oil pressure has built up, start the engine and allow it to run until oil is ejected from the oil filter bleed screw orifice. The oil will be forced out under some considerable pressure so be prepared to turn the engine off, and have a suitable rag handy. Fit and tighten the bleed screw, and then check the oil level in the tank, replenishing as necessary.

2 Adjusting the valve clearances

To gain access to the rocker covers and rockers the petrol tank must be removed as described in Chapter 2, Section 2. Remove the sump guard (where fitted). Slacken and remove all the screws which retain the left-hand engine cover and remove the cover. Remove the spark plug.

Remove the rocker covers, each of which is retained by two screws. Rotate the engine so that the piston is at TDC on the compression stroke; in this position both valves will be closed and the T mark scribed on the rotor will be in alignment with the index pointer projecting from the crankcase.

Using a feeler gauge of suitable size, check the clearance between each rocker arm and valve stem end. The clearances should be within the following ranges. Note that the valve clearances must be checked and adjusted when the engine is quite cold.

Valve clearances

	SR500 model	**All other models**
Inlet	0.10 mm	0.07-0.12 mm
	(0.004 in)	(0.003-0.005 in)
Exhaust	0.15 mm	0.12-0.17 mm
	(0.006 in)	(0.005-0.007 in)

If the clearance on either valve is incorrect, slacken the locknut on the adjuster screw and turn the screw in or out until the feeler gauge is a light sliding fit between the surfaces. Hold the adjuster still and then tighten fully the locknut. Re-check once the locknut has been tightened. Where a clearance range is given, the smallest sizes in the range should be adopted when resetting the clearance.

Before reassembling the engine components and replacing the petrol tank refer to the next maintenance heading and carry out the initial steps in the cam chain tensioning procedure.

3 Checking and adjusting the cam chain tension

Rotate the engine forwards (anti-clockwise viewed from left-hand side) so that the slack in the cam chain is placed in the rear run of the chain. The crankshaft should be stopped so that the piston is at TDC on the compression stroke. Remove the cover from the chain tensioner assembly to the rear of the cylinder barrel and check that the end of the tensioner plunger is flush with the outer face of the small slotted adjuster screw head. If the plunger is not correctly positioned, loosen the adjuster locknut (the large hexagon) and screw the adjuster in or out as necessary. Tighten the locknut fully. Refit the petrol tank, sparking plug, rocker covers and the inspection plug and start the engine. With the engine running at tick-over speed, the tensioner plunger should be seen to move in and out slightly, indicating that the tensioner is correctly set. If no movement occurs, the tensioner is too tight; slacken the locknut and unscrew the adjuster a few degrees. Tighten the locknut and check again for movement. If all is correct, fit and tighten the adjuster cap.

Grease the swinging arm bearings via the nipple

Remove sump drain plug and ...

... oil tank (frame tube) plug to drain

Adjust the valve clearance so that the gauge is a sliding fit

4   Adjusting the contact breaker – except SR500 models

To gain access to the contact breaker assembly, it is necessary to detach the cover secured on the primary drive cover by two screws.

Rotate the engine slowly by means of the kickstarter until the points are in the fully-open position. Examine the faces of the contacts; if they are pitted or burnt, it will be necessary to remove them for further attention, as described in Chapter 3, Section 5.

Adjustment is carried out by slackening the two screws which retain the fixed contact breaker point and inserting a screwdriver in the adjusting slot provided. Turn in the appropriate direction until the gap is within the range 0.3-0.4 mm (0.012-0.016 inch) and then retighten the two retaining screws. It is imperative that the points are open FULLY whilst this adjustment is made, or a false reading will result.

Before replacing the cover and gasket, place a slight smear of grease on the contact breaker cam and a few drops of thin oil on the felt wick which lubricates the surface of the cam. Do not over lubricate the wick, because excess oil may find its way onto the points faces, causing ignition malfunction.

5   Checking the ignition timing – except SR500 models

The ignition timing should be checked, and if necessary adjusted, after adjustment of the contact breaker has been completed. Refer to Chapter 3 Section 9 for the correct procedure to follow for both static and stroboscopic timing.

SR500 models only

The SR500 models are fitted with a CDI ignition system which requires checking much less frequently than the contact breaker system fitted to all the other models. Refer to Chapter 3, Section 9 for the relevant details when the half-yearly or 4,000 mile service falls due.

6   Air filter cleaning – except SR500 models

After removal of the frame right-hand side cover, to gain access to the air filter box the air box cover may be removed; it is retained by three screws. The filter element is supported on a frame which is fitted by detachable plates at the front and rear. Grasp the two plates and slide the complete assembly from place. Detach the two plates and remove the element from the frame. The element should be cleaned thoroughly in petrol. Squeeze out the element to remove as much petrol as possible and then leave the element for a short while for the remainder of the petrol to evaporate. Do not wring out the sponge as this will cause damage to the fabric, leading to the need for early replacement. Reimpregnate the element with SAE 30 engine oil and then squeeze it gently to remove excess oil. The element should be wet but not dripping. If the sponge becomes damaged or hardens with age, it should be renewed.

Refit the element by reversing the dismantling procedure. The faces of the element frame should be coated with grease to ensure an airtight seal between them and the support plates. Check also that the sealing ring on the front plate is in good condition and seating correctly.

Never run the machine without the element or with the air filter disconnected, otherwise the weak mixture that results will cause overheating which may lead to seizure or piston damage.

Pull out the air filter plates and element simultaneously

7   Adjusting the carburettor

The procedure which follows may be disregarded if the engine tick-over is correct and no roughness is evident at idle speeds.

Adjustment of the mixture for a regular tick-over speed and adjustment of the tick-over speed itself should be carried out only after the engine has been allowed to reach normal temperature. If the adjustments are made when the engine is cold the results will be incorrect during normal operation.

Screw in the pilot screw until it seats lightly and then screw it out exactly the amount specified below for each model. It should be noted, however, that on TT500E, XT500E models and SR500 models the pilot screw is fitted with a limiter cap. On adjustment at the factory the cap is fitted to prevent injudicious over-adjustment of the screw. Where the screw limiter is fitted do not move the screw at this stage.

TT500C and XT500D	1¼ turns out
TT500D and XT500C	1¾ turns out

Start the engine and allow it to tick-over. Turn the pilot adjuster screw in or out a small amount until the highest rpm is found. On those carburettors fitted with a limiter cap, the screw should be moved only within the confines of the limit stops.

Using the throttle stop screw with the knurled head adjust the tick-over speed to the following rpm for each model.

SR500, TT500C and XT500D	1100 rpm
TT500D and E and XT500E	1200 rpm

Carry out the pilot screw adjustment once more and then, if required, repeat the tick-over adjustment operation.

On those machines not fitted with a tachometer and where an ancillary tachometer is not available for testing purposes, the tick-over should be set by ear.

8   Sparking plug cleaning and re-setting

Remove the sparking plug and clean it, using a wire brush. Additionally, clean the electrode faces using fine emery paper or a fine file. Re-set the points gap using a 0.7mm (0.028 in) feeler gauge to determine when the gap is correct. Bend the outer electrode only, do not bend the central electrode because this will damage the ceramic insulator. Before fitting the plug, smear the threads with a graphited grease; this will aid future removal.

Half-yearly or every 4000 miles (6000 kilometres)

When this point is reached, all the foregoing checks should be carried out first, working from the weekly/200 mile service right through to the quarterly/2000 mile list.

1   Oil filter change

At every second engine/transmission oil change the paper element oil filter should be removed and discarded, a new element being fitted in its place. In addition to this, the gauze oil strainer fitted to the oil feed union at the lower end of the front frame tube, and the screen fitted in the crankcase sump should be removed and cleaned. These operations should, of course, be carried out after draining the engine/transmission oil in the normal manner.

The main oil filter is contained within a chamber in the right-hand engine cover, which is closed by a small cover retained by three screws. The lower screw also acts as a drain plug, so that the small amount of oil in the chamber may be drained prior to removal of the cover. Slacken the three screws and then remove the lower screw and drain the chamber. Remove the remaining screws and lift off the cover. Note the large 'O' ring, which seals the cover and also the smaller 'O' ring which seals the drain hole in the casing. Lift out the oil filter and fit a new filter of the same type. The filter should be fitted with the by-pass valve flap towards the outside. Replace the cover and tighten the screws evenly.

To gain access to the two gauze filters, the sump guard must be removed. This is held by four or six bolts, depending on the model to which it is fitted. Loosen evenly and remove the screws holding the sump cover in place. Lower the cover from place and pull out the sheet gauze screen. To remove the oil feed union screen, unscrew the gland nut holding the pipe to the union at the bottom of the frame front downtube. Unscrew the union and withdraw it, together with the filter column.

Both filters should be cleaned thoroughly in petrol and then dried, before they are refitted. Check that the sealing washer on the feed union is in good condition and that the sump cover gasket is not damaged. Replace either component if suspect.

2   Front fork oil change

Support the machine on blocks placed below the crankcase so that the front wheel is clear of the ground. On SR500 models, prise the rubber cap from the top of each fork stanchion, and using a suitable socket screw remove the recessed plugs. On all other models unscrew the chromed bolts from the top of each fork stanchion. Each leg is fitted with a drain plug above and to the rear of the wheel spindle; remove the plugs and allow the oil to drain into suitable containers. When most of the oil has drained, pump the fork legs up and down to expel any remaining fluid. Refit and tighten the drain plugs.

Refill each fork leg with the following type and quantity of damping fluid.

	TT500C and XT500C	TT500D and E, and XT500D and E	SR500
Oil type	SAE20 fork oil or engine oil	SAE10W/30	SAE10 fork oil
Oil capacity	217 cc (7.3/6.1 US/Imp fl oz)	223 cc (7.5/6.3 US/Imp fl oz)	182 cc (6.1/5.1 US/Imp fl oz)

Replace the fork top plugs of whichever type is fitted. Note the 'O' ring fitted to each plug on TT and XT500 models. These should be renewed if they have perished or broken.

3 Ignition timing check – SR500 models

The CDI ignition fitted to the SR500 models requires that the ignition timing be checked less often than on a machine fitted with normal contact breaker equipment. This is because there are no mechanical parts to wear and so reduce the accuracy of the timing. The likelihood of adjustments being necessary at this check is small; nevertheless the check should be carried out to ensure that no alteration has occured. Refer to Chapter 3, Section 9 for the correct procedure.

4 Instrument drive cable lubrication

Lubricate the speedometer and tachometer drive cables by detaching the cable at the driving end and withdrawing the inner cable. High melting point or graphited grease should be used, taking care not to grease the last six inches of the cable at the instrument head end. If this precaution is overlooked, grease will work into the instrument head and immobilise the movement.

5 Steering head bearings

Check the steering head bearings for play by applying the front brake hard and seeing whether there is any movement as the machine is rocked backwards and forwards. If adjustment is necessary, slacken the pinch bolt through the upper fork yoke and also the yoke centre bolt. Make the adjustment by tightening or loosening the slotted adjuster ring fitted to the steering stem immediately below the upper fork yoke. A 'C' spanner should be used to turn the ring.

Retighten the pinch bolt, and re-check. Do not overtighten, or the machine will show a tendency to roll at low speeds. Do not confuse movement at the steering head with worn fork sliders.

Fit the oil filter as shown; note the two O rings

Oil tank filter is integral with feed union

Sump filter is a slot fit in the sump cover

Remove screw plug to drain fork leg

Refill with correct quantity and specification of damping fluid

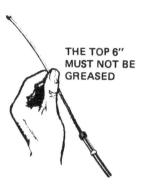

THE TOP 6"
MUST NOT BE
GREASED

Greasing the speedometer cable

Yearly, or every 8000 miles (12000 kilometres)

The yearly/8000 mile check should be considered as a minor overhaul. In addition to carrying out all the tasks listed under the preceding routine maintenance sections, the following should be attended to:

1 Wheel bearings

The wheels should be removed so that access can be made to the bearings for removal, examination and relubrication. Refer to Chapter 5, Section 12 for the front wheel and Section 17 for the rear wheel.

2 Brake linings – drum brakes

After removal of the wheel(s) the drum brake(s) should be examined and renovated as described in Chapter 5, Section 5.

3 Steering head bearings

Dismantle the front fork assembly so that the steering head bearings may be inspected for wear and re-lubricated. Refer to Chapter 5, Sections 2, 3 and 6.

General maintenance adjustments and inspection

1 Clutch adjustment

The intervals at which the clutch should be adjusted will depend on the style of riding and the conditions under which the machine is used.

Adjust the clutch in two stages to ensure smooth operation as follows:

Slacken the cable adjuster so that there is plenty of free play in the cable. On inspection it will be seen that the clutch operating arm has a small scribed line marked near the cable anchor, and that a small projection protrudes from the crankcase adjacent to the arm. Push the operating arm forwards until all lost-motion is taken up and it can be felt that the operating cam has abutted against the pushrod end. In this position the scribed line and the projection should be in alignment. If this is not the case, remove the final drive sprocket cover and slacken the locknut on the clutch mechanism adjuster in the gearbox left-hand wall. Turn the adjuster screw in or out until alignment is correct. Tighten the locknut and refit the cover. Adjust the free play in the cable so that there is approximately 12 mm ($\frac{1}{2}$ in) free play, at the handlebar lever ball end, before the clutch begins to lift.

Slacken the locknut and rotate adjuster to set clutch arm position

2 Checking brake pad wear

Brake pad wear depends largely on the conditions under which the machine is ridden and at what speed. It is difficult therefore to give precise inspection intervals, but it follows that pad wear should be checked more frequently on a hard ridden machine.

The condition of each pad can be checked easily whilst still in situ on the machine. The pads have a red groove around their outer periphery which can be seen if the small inspection cover in each caliper is lifted. If wear has reduced either or both pads in one caliper down to the red line the pads should be renewed as a pair.

Each set of pads may be removed with the wheel in place by detaching the caliper. There is no need to disconnect the brake hose. Unscrew the two bolts which pass through the fork leg into the caliper support bracket. Lift the assembly off the brake disc and remove the single bolt which holds the caliper unit to the bracket. Unscrew the single crosshead screw from the inner face of the caliper, noting that this screw acts as a locator for the pads. Pull the support bracket from the main unit and lift out the pads. Note the various shims and their positions. The anti-chatter spring fitted between the pad and piston is fitted with the arrow pointing in the direction of wheel rotation.

Fit new pads by reversing the dismantling procedure. If difficulty is encountered when fitting the caliper over the disc, due to the reduced distance between the pads, use a wooden lever to push the piston side pad inwards.

3 Cleaning the machine

Keeping the motorcycle clean should be considered an important part of the routine maintenance of the machine. Not only are developing faults more readily seen on a clean machine, but the rectification of such faults is less laborious if the components in question are already in a fit state to be attended to.

After removing all surface dirt with a rag or sponge which is washed frequently in clean water, the machine should be allowed to dry thoroughly. Application of car polish or wax to the cycle parts will give a good finish, particularly if the machine receives this attention at regular intervals.

If the machine has been used off-road, it will prove much more difficult to clean and it is advisable to give it a good hose down first. Take care to keep water out of the air filter and the electrics or the carburettor since these are the more vulnerable items. Allow the machine to dry thoroughly before wiping over with a clean rag and then polishing.

The plated parts should require only a wipe with a damp rag, but if they are badly corroded, as may occur during the winter when the roads are salted, it is permissible to use one of the proprietary chrome cleaners. These often have an oily base which will help to prevent corrosion from recurring.

If the engine parts are particularly oily, use a cleaning compound such as Gunk or Jizer. Apply the compound whilst the parts are dry and work it in with a brush so that it has an opportunity to penetrate and soak into the film of oil and grease. Finish off by washing down liberally, taking care that water does not enter the carburettors, air cleaners or the electrics. If desired, the now clean alloy parts can be enhanced still further when they are dry by using a special polish such as Solvol Autosol. This will restore the full lustre.

If possible, the machine should be wiped down immediately after it has been used in the wet, so that it is not garaged under damp conditions that will promote rusting. Make sure the chain is wiped and re-oiled, to prevent water from entering the rollers and causing harshness with an accompanying rapid rate of wear. Remember there is less chance of water entering the control cables and causing stiffness if they are lubricated regularly as described in the Routine Maintenance section.

Quick glance maintenance adjustments and capacities

Engine oil capacity

Dry	2.4 litre (5.0/4.2 US/Imp pints)
At oil change	2.0 litre (4.2/3.5 US/Imp pints)
At oil and filter change	2.1 litre (4.4/3.7 US/Imp pints)

Front forks

Damping fluid type	
SR500 model	SAE 10 fork oil
All other models	SAE 20 fork oil or SAE 10W/30 engine oil
Capacity (per leg)	
TT500C and XT500C	217 cc (7.3/6.1 US/Imp fl oz)
TT500D and E, XT500D and E	223 cc (7.5/6.3 US/Imp fl oz)
SR500	182 cc (6.1/5.1 US/Imp fl oz)

Contact breaker gap	0.3 – 0.4 mm (0.012 – 0.016 in)

Sparking plug gap	0.7 – 0.8 mm (0.028 – 0.031 in)

Valve clearances (Cold)

	Inlet	Exhaust
SR500	0.10 mm (0.004 in)	0.15 mm (0.006 in)
All others	0.07 – 0.12 mm (0.003 – 0.005 in)	0.12 – 0.17 mm (0.005 – 0.007 in)

Tyre pressures

	TT500C,D and E	XT500C,D and E	SR500
Front:			
Off road	13 psi	13 psi	—
*On road	—	18 psi	26 psi
Rear:			
Off road	16 psi	16 psi	—
*On road	—	21 psi	28 psi

*When carrying a passenger or travelling at continuous high speed increase the front tyre pressure by 2 – 3 psi and the rear tyre pressure by 3 – 4 psi

Recommended lubricants

Component	Lubricant
Engine/gearbox	SAE 20W/40 or SAE 20W/50 motor oil
Front forks: SR500 model	SAE 10 fork oil
All other models	SAE 20 fork oil or SAE 10W/30 engine oil
Wheel bearings	High melting point grease
Hydraulic brakes	DOT 3 (USA), SAE J1703 (UK)

Working conditions and tools

When a major overhaul is contemplated, it is important that a clean, well-lit working space is available, equipped with a workbench and vice, and with space for laying out or storing the dismantled assemblies in an orderly manner where they are unlikely to be disturbed. The use of a good workshop will give the satisfaction of work done in comfort and without haste, where there is little chance of the machine being dismantled and reassembled in anything other than clean surroundings. Unfortunately, these ideal working conditions are not always practicable and under these latter circumstances when improvisation is called for, extra care and time will be needed.

The other essential requirement is a comprehensive set of good quality tools. Quality is of prime importance since cheap tools will prove expensive in the long run if they slip or break when in use, causing personal injury or expensive damage to the component being worked on. A good quality tool will last a long time, and more than justify the cost.

For practically all tools, a tool factor is the best source since he will have a very comprehensive range compared with the average garage or accessory shop. Having said that, accessory shops often offer excellent quality tools at discount prices, so it pays to shop around. There are plenty of tools around at reasonable prices, but always aim to purchase items which meet the relevant national safety standards. If in doubt, seek the advice of the shop proprietor or manager before making a purchase.

The basis of any tool kit is a set of open-ended spanners, which can be used on almost any part of the machine to which there is reasonable access. A set of ring spanners makes a useful addition, since they can be used on nuts that are very tight or where access is restricted. Where the cost has to be kept within reasonable bounds, a compromise can be effected with a set of combination spanners – open-ended at one end and having a ring of the same size on the other end. Socket spanners may also be considered a good investment, a basic $3/8$ in or $1/2$ in drive kit comprising a ratchet handle and a small number of socket heads, if money is limited. Additional sockets can be purchased, as and when they are required. Provided they are slim in profile, sockets will reach nuts or bolts that are deeply recessed. When purchasing spanners of any kind, make sure the correct size standard is purchased. Almost all machines manufactured outside the UK and the USA have metric nuts and bolts, whilst those produced in Britain have BSF or BSW sizes. The standard used in USA is AF, which is also found on some of the later British machines. Others tools that should be included in the kit are a range of crosshead screwdrivers, a pair of pliers and a hammer.

When considering the purchase of tools, it should be remembered that by carrying out the work oneself, a large proportion of the normal repair cost, made up by labour charges, will be saved. The economy made on even a minor overhaul will go a long way towards the improvement of a toolkit.

In addition to the basic tool kit, certain additional tools can prove invaluable when they are close to hand, to help speed up a multitude of repetitive jobs. For example, an impact screwdriver will ease the removal of screws that have been tightened by a similar tool, during assembly, without a risk of damaging the screw heads. And, of course, it can be used again to retighten the screws, to ensure an oil or airtight seal results. Circlip pliers have their uses too, since gear pinions, shafts and similar components are frequently retained by circlips that are not too easily displaced by a screwdriver. There are two types of circlip pliers, one for internal and one for external circlips. They may also have straight or right-angled jaws.

One of the most useful of all tools is the torque wrench, a form of spanner that can be adjusted to slip when a measured amount of force is applied to any bolt or nut. Torque wrench settings are given in almost every modern workshop or service manual, where the extent to which a complex component, such as a cylinder head, can be tightened without fear of distortion or leakage. The tightening of bearing caps is yet another example. Overtightening will stretch or even break bolts, necessitating extra work to extract the broken portions.

As may be expected, the more sophisticated the machine, the greater is the number of tools likely to be required if it is to be kept in first class condition by the home mechanic. Unfortunately there are certain jobs which cannot be accomplished successfully without the correct equipment and although there is invariably a specialist who will undertake the work for a fee, the home mechanic will have to dig more deeply in his pocket for the purchase of similar equipment if he does not wish to employ the services of others. Here a word of caution is necessary, since some of these jobs are best left to the expert. Although an electrical multimeter of the AVO type will prove helpful in tracing electrical faults, in inexperienced hands it may irrevocably damage some of the electrical components if a test current is passed through them in the wrong direction. This can apply to the synchronisation of twin or multiple carburettors too, where a certain amount of expertise is needed when setting them up with vacuum gauges. These are, however, exceptions. Some instruments, such as a strobe lamp, are virtually essential when checking the timing of a machine powered by CDI ignition system. In short, do not purchase any of these special items unless you have the experience to use them correctly.

Although this manual shows how components can be removed and replaced without the use of special service tools (unless absolutely essential), it is worthwhile giving consideration to the purchase of the more commonly used tools if the machine is regarded as a long term purchase. Whilst the alternative methods suggested will remove and replace parts without risk of damage, the use of the special tools recommended and sold by the manufacturer will invariably save time.

Chapter 1 Engine, clutch and gearbox

Refer to Chapter 7 for information relating to the 1979 to 1983 models

Contents

Specifications

Engine

Type .... Single cylinder, overhead camshaft, four-stroke
Bore .... 87 mm (3.425 in)
Stroke .... 84 mm (3.307 in)
Capacity .... 499 cc (30.45 cu in)
Compression ratio .... 9 : 1

Cylinder barrel

Type .... Aluminium alloy with cast iron liner
Standard bore .... 87.00-87.02 mm (3.4252-3.4260 in)
Wear limit .... 87.1 mm (3.429 in)
Taper limit .... 0.05 mm (0.0020 in)
Ovality limit .... 0.01 mm (0.0004 in)
Cylinder bore/piston clearance .... 0.050-0.055 mm (0.0020-0.0022 in)
Wear limit .... 0.1 mm (0.004 in)

Safety first!

Professional motor mechanics are trained in safe working procedures. However enthusiastic you may be about getting on with the job in hand, do take the time to ensure that your safety is not put at risk. A moment's lack of attention can result in an accident, as can failure to observe certain elementary precautions.

There will always be new ways of having accidents, and the following points do not pretend to be a comprehensive list of all dangers; they are intended rather to make you aware of the risks and to encourage a safety-conscious approach to all work you carry out on your vehicle.

Essential DOs and DON'Ts

DON'T start the engine without first ascertaining that the transmission is in neutral.

DON'T suddenly remove the filler cap from a hot cooling system – cover it with a cloth and release the pressure gradually first, or you may get scalded by escaping coolant.

DON'T attempt to drain oil until you are sure it has cooled sufficiently to avoid scalding you.

DON'T grasp any part of the engine, exhaust or silencer without first ascertaining that it is sufficiently cool to avoid burning you.

DON'T allow brake fluid or antifreeze to contact the machine's paintwork or plastic components.

DON'T syphon toxic liquids such as fuel, brake fluid or antifreeze by mouth, or allow them to remain on your skin.

DON'T inhale dust – it may be injurious to health (see *Asbestos* heading).

DON'T allow any spilt oil or grease to remain on the floor – wipe it up straight away, before someone slips on it.

DON'T use ill-fitting spanners or other tools which may slip and cause injury.

DON'T attempt to lift a heavy component which may be beyond your capability – get assistance.

DON'T rush to finish a job, or take unverified short cuts.

DON'T allow children or animals in or around an unattended vehicle.

DON'T inflate a tyre to a pressure above the recommended maximum. Apart from overstressing the carcase and wheel rim, in extreme cases the tyre may blow off forcibly.

DO ensure that the machine is supported securely at all times. This is especially important when the machine is blocked up to aid wheel or fork removal.

DO take care when attempting to slacken a stubborn nut or bolt. It is generally better to pull on a spanner, rather than push, so that if slippage occurs you fall away from the machine rather than on to it.

DO wear eye protection when using power tools such as drill, sander, bench grinder etc.

DO use a barrier cream on your hands prior to undertaking dirty jobs – it will protect your skin from infection as well as making the dirt easier to remove afterwards; but make sure your hands aren't left slippery. Note that long-term contact with used engine oil can be a health hazard.

DO keep loose clothing (cuffs, tie etc) and long hair well out of the way of moving mechanical parts.

DO remove rings, wristwatch etc, before working on the vehicle – especially the electrical system.

DO keep your work area tidy – it is only too easy to fall over articles left lying around.

DO exercise caution when compressing springs for removal or installation. Ensure that the tension is applied and released in a controlled manner, using suitable tools which preclude the possibility of the spring escaping violently.

DO ensure that any lifting tackle used has a safe working load rating adequate for the job.

DO get someone to check periodically that all is well, when working alone on the vehicle.

DO carry out work in a logical sequence and check that everything is correctly assembled and tightened afterwards.

DO remember that your vehicle's safety affects that of yourself and others. If in doubt on any point, get specialist advice.

IF, in spite of following these precautions, you are unfortunate enough to injure yourself, seek medical attention as soon as possible.

Asbestos

Certain friction, insulating, sealing, and other products – such as brake linings, clutch linings, gaskets, etc – contain asbestos. *Extreme care must be taken to avoid inhalation of dust from such products since it is hazardous to health.* If in doubt, assume that they *do* contain asbestos.

Fire

Remember at all times that petrol (gasoline) is highly flammable. Never smoke, or have any kind of naked flame around, when working on the vehicle. But the risk does not end there – a spark caused by an electrical short-circuit, by two metal surfaces contacting each other, by careless use of tools, or even by static electricity built up in your body under certain conditions, can ignite petrol vapour, which in a confined space is highly explosive.

Always disconnect the battery earth (ground) terminal before working on any part of the fuel or electrical system, and never risk spilling fuel on to a hot engine or exhaust.

It is recommended that a fire extinguisher of a type suitable for fuel and electrical fires is kept handy in the garage or workplace at all times. Never try to extinguish a fuel or electrical fire with water.

Note: *Any reference to a 'torch' appearing in this manual should always be taken to mean a hand-held battery-operated electric lamp or flashlight. It does **not** mean a welding/gas torch or blowlamp.*

Fumes

Certain fumes are highly toxic and can quickly cause unconsciousness and even death if inhaled to any extent. Petrol (gasoline) vapour comes into this category, as do the vapours from certain solvents such as trichloroethylene. Any draining or pouring of such volatile fluids should be done in a well ventilated area.

When using cleaning fluids and solvents, read the instructions carefully. Never use materials from unmarked containers – they may give off poisonous vapours.

Never run the engine of a motor vehicle in an enclosed space such as a garage. Exhaust fumes contain carbon monoxide which is extremely poisonous; if you need to run the engine, always do so in the open air or at least have the rear of the vehicle outside the workplace.

The battery

Never cause a spark, or allow a naked light, near the vehicle's battery. It will normally be giving off a certain amount of hydrogen gas, which is highly explosive.

Always disconnect the battery earth (ground) terminal before working on the fuel or electrical systems.

If possible, loosen the filler plugs or cover when charging the battery from an external source. Do not charge at an excessive rate or the battery may burst.

Take care when topping up and when carrying the battery. The acid electrolyte, even when diluted, is very corrosive and should not be allowed to contact the eyes or skin.

If you ever need to prepare electrolyte yourself, always add the acid slowly to the water, and never the other way round. Protect against splashes by wearing rubber gloves and goggles.

Mains electricity and electrical equipment

When using an electric power tool, inspection light etc, always ensure that the appliance is correctly connected to its plug and that, where necessary, it is properly earthed (grounded). Do not use such appliances in damp conditions and, again, beware of creating a spark or applying excessive heat in the vicinity of fuel or fuel vapour. Also ensure that the appliances meet the relevant national safety standards.

Ignition HT voltage

A severe electric shock can result from touching certain parts of the ignition system, such as the HT leads, when the engine is running or being cranked, particularly if components are damp or the insulation is defective. Where an electronic ignition system is fitted, the HT voltage is much higher and could prove fatal.

Piston and rings

Piston oversizes:
    1st oversize . . . .   87.25 mm
    2nd oversize . . . .   87.50 mm
    3rd oversize . . . .   87.75 mm
    4th oversize . . . .   88.00 mm
Ring end gap:
    Top and second ring . . . .   0.3-0.5 mm (0.012-0.020 in)
    Wear limit . . . .   0.8 mm (0.031 in)
    Oil control ring . . . .   0.2-0.9 mm (0.008-0.035 in)
    Wear limit . . . .   1.0 mm (0.039 in)
Side clearance:
    Top ring . . . .   0.04-0.08 mm (0.0016-0.0031 in)
    Wear limit . . . .   0.15 mm (0.006 in)
    Second ring . . . .   0.03-0.07 mm (0.0012-0.0028 in)
    Wear limit . . . .   0.15 mm (0.006 in)

Valves and valve springs

Valve seat angle . . . .   45°
Valve stem diameter:
    Inlet . . . .   7.97-7.99 mm (0.3138-0.3146 in)
    Exhaust . . . .   7.96-7.97 mm (0.3134-0.3136 in)
Valve stem/guide clearance:
    Inlet . . . .   0.02-0.04 mm (0.0008-0.0016 in)
    Wear limit . . . .   0.08 mm (0.003 in)
    Exhaust . . . .   0.04-0.06 mm (0.0016-0.0024 in)
    Wear limit . . . .   0.1 mm (0.0040 in)
Valve spring free length:
    Inner . . . .   45.25 mm (1.7815 in)
    Wear limit . . . .   43.9 mm (1.7283 in)
    Outer . . . .   45.15 mm (1.7776 in)
    Wear limit . . . .   43.8 mm (1.7244 in)

Valve clearance (cold)

	XT and TT models	SR500 models
Inlet	0.07-0.12 mm (0.003-0.005in)	0.10 mm (0.004 in)
Exhaust	0.12-0.17 mm (0.005-0.007 in)	0.15 mm (0.006 in)

Valve timing

Inlet opens . . . .   44° BTDC
Inlet closes . . . .   68° ABDC
Exhaust opens . . . .   76° BBDC
Exhaust closes . . . .   36° ATDC

Camshaft and rockers

Cam lobe height:
    Inlet . . . .   39.18-39.28 mm (1.5425-1.5465 in)
    Wear limit . . . .   39.08 mm (1.5386 in)
    Exhaust . . . .   39.20-39.27 mm (1.5433-1.5473 in)
    Wear limit . . . .   39.10 mm (1.5394 in)
Rocker spindle diameter . . . .   11.98-11.99 mm (0.4717-0.4720 in)
Wear limit . . . .   11.96 mm (0.4709 in)
Spindle/bearing clearance . . . .   0.01-0.04 mm (0.0004-0.0016 in)
Wear limit . . . .   0.11 mm (0.0043 in)

Crankshaft

Maximum run-out . . . .   0.03 mm (0.0012 in)
Connecting rod side clearance . . . .   0.35-0.65 mm (0.0138-0.0256 in)
Rod deflection . . . .   0.8-1.0 mm (0.0315-0.0394 in)

Clutch

Type . . . .   Wet, multi-plate
No.of plates:
    Plain   7
    Inserted   8
No. of springs   6
Inserted plate thickness . . . .   2.8 mm (0.110 in)
Wear limit . . . .   2.5 mm (0.098 in)
Plain plate maximum warpage . . . .   0.05 mm (0.002 in)
Spring free length . . . .   41.2 mm (1.622 in)
Wear limit . . . .   40.0 mm (1.575 in)

Gearbox

Type	5-speed, constant mesh
Gear ratios:	
1st	2.357 : 1
2nd	1.555 : 1
3rd	1.190 : 1
4th	0.916 : 1
5th	0.777 : 1
Primary drive ratio	2.566 : 1
Final drive ratio:	
TT500C	3.466 : 1
TT500D	3.334 : 1
TT500E, XT500C, D and E and SR500	2.750 : 1

Main torque wrench settings

Cylinder head:	
10 mm bolt	1.7-2.3 kgf m (12.0-17.0 lbf ft)
10 mm nut	3.5-4.0 kgf m (25.0-29.0 lbf ft)
8 mm screw	1.0-1.5 kgf m (7.0-11.0 lbf ft)
8 mm nut	1.8-2.2 kgf m (13.0-16.0 lbf ft)
6 mm screw	0.8-1.2 kgf m (5.8-8.7 lbf ft)
Cylinder barrel	
10 mm nut	3.5-4.0 kgf m (25.0-29.0 lbf ft)
6 mm screw	0.8-1.2 kgf m (5.8-8.7 lbf ft)
Flywheel generator rotor nut	8.0 kgf m (58.0 lbf ft)
Valve clearance adjuster locknut	2.7 kgf m (19.5 lbf ft)
ATU bolt	0.8 kgf m (5.8 lbf ft)
Primary drive gear nut	6.0 kgf m (50.5 lbf ft)
Clutch centre nut	6.0 kgf m (50.5 lbf ft)

1 General description

The engine fitted to the Yamaha TT, XT and SR500 models is of the overhead camshaft type in which the valve gear is actuated by means of rockers that bear direct on the camshaft. The camshaft is driven by an endless chain which passes through the right-hand side of the cylinder barrel and head.

All main castings are in aluminium alloy, to save weight. A few of the castings, such as the crankcase covers, are in magnesium alloy, which is even lighter in weight although more expensive to manufacture and machine. The crankshaft is of the built-up type consisting of two full flywheels and separate mainshafts, and running on two journal ball bearings of generous proportions. The connecting rod big-end is of the caged needle roller type, having eighteen 4 mm rollers. The crankcase castings contain also the gearbox components and in consequence when attention is required to either the engine or the gearbox, the engine unit must be removed from the frame and the crankcase separated. The crankcase halves separate in the traditional manner in a vertical plane. A flywheel magneto generator is mounted on the left-hand side of the engine; the clutch is on the right-hand side. Both are fully enclosed behind detachable side covers. An upswept exhaust system is carried on the side of the machine, terminating in a large capacity silencer that gives a very quiet exhaust note. There is no electric start; the engine is started by a kickstarter mounted on the right-hand side of the machine, in the conventional manner.

Lubrication is effected by means of a dry sump system in which the oil is contained within the frame top and front downtube and delivered and returned from the oil pump by external hoses. A double trochoid oil pump, driven by a single shaft, is used to transmit the oil. The pump incorporates two completely separate units, one of which delivers oil under pressure to the working parts of the engine. The circulated oil returns to the sump where it is picked-up by the scavenge pump unit and returned to the oil reservoir in the frame. A full flow oil filter and two gauze screens are included in the system to prevent circulation of contaminants.

2 Operations with engine/gearbox in frame

It is not necessary to remove the engine unit from the frame if the following items require attention:

1 Clutch assembly
2 Oil pump and oil filter
3 Flywheel generator
4 Contact breaker and auto-advance assembly

If several items require attention, it is often more convenient to remove the complete engine unit from the frame, in order to gain better access. This operation will take approximately one hour to complete.

3 Operations with engine/gearbox removed

Certain operations can be accomplished only if the complete engine unit is removed from the frame. This is because it is necessary to separate the crankcases to gain access to the parts concerned, or because there is insufficient clearance to withdraw parts after they have been slackened or freed from their normal location. These operations include:

1 Removal and replacement of the main bearings
2 Removal and replacement of the crankshaft assembly
3 Removal and replacement of the gear cluster, selectors and gearbox main bearings
4 Removal and replacement of the cylinder head, cylinder barrel and piston

4 Method of engine/gearbox removal

As mentioned previously, the engine and gearbox are built in unit and it is necessary to remove the unit complete in order to gain access to either component. Separation is accomplished after the engine/gearbox unit has been removed from the frame and refitting cannot take place until the crankcases have been reassembled. When the crankcases are separated, the gearbox internals will also be exposed. There is no means of working on the engine without disturbing the gearbox, or on the gearbox without first dismantling the engine.

5   Removing the engine/gearbox unit

1   With the machine supported on the side-stand, remove the sump guard. This is held in place either by four small bolts (C models) or six bolts (D models and later). The SR500 has only a small cover held by four bolts. Place the machine astride a stout wooden box or some similar form of packing so that it will stand unsupported on level ground. This procedure is necessary because the machine has no centre stand. Position the machine so that access is available to the drain plug on the underside of the crankcase, and to the drain plug at the lower end of the frame front downtube.

2   Position a container that will hold at least 2¼ litres (4 pints) under the crankcase drain plug and remove the latter so that the engine oil will drain off. To aid draining, unscrew the small bleed plug from the top of the oil filter housing on the right-hand side of the engine. Draining the oil from the engine is best carried out when the engine is warm; the oil will be thinner and so drain more quickly and more completely. When the oil has finished draining, replace and tighten the drain plug, making sure the sealing washer is in good condition.

3   Move the container with the oil drainings to the underside of the frame front tube. Remove the filler cap and then the drain plug. When the oil has drained, disconnect the oil feed pipe from the union at the base of the downtube and unscrew and withdraw the union, complete with the integral filter column. A small additional quantity of oil will drain off when the plug complete with filter gauze is unscrewed.

4   Detach the side panel (frame cover) from each side of the machine. The panel on the left-hand side of the machine (except TT500 models) is secured by a lock at the lower edge and by two projections at the upper edge. The right-hand side panel, (and both panels on TT models) is held in a similar manner, the lock being substituted by a screw or slotted Dzus fastener. Remove the dualseat after unscrewing the two bolts which pass through lugs in the frame and into the base of the seat. Release the battery retaining strap and disconnect both battery leads either at the snap connectors, or by removing the terminal screws. If it is anticipated that the machine is to be out of commission for an extended length of time, the battery should be given a refresher charge at regular intervals. See Chapter 6.

5   The method required for removal of the exhaust system differs radically for each model type. On TT500C and XT500C machines the complete assembly may be removed as follows. Unscrew the two mounting bolts holding the right-hand rear suspension unit in place. Temporarily support the weight of the machine until the unit has been pulled from place. Unscrew the two knurled socket nuts which secure the exhaust pipe flange at the exhaust port. The exhaust system is supported by rubber mountings through which pass the single forward securing stud and the two rear securing bolts. After removal of these, the complete unit may be lifted away. Note the sequence of cups, rubbers and spacers which comprise the lowest mounting assembly. The exhaust system fitted to TT500D and E and XT500D and E models does not require removal as a complete unit. Sufficient access may be gained for engine removal by detaching only the exhaust pipe proper. Loosen the clamp bolt at the exhaust pipe/silencer joint and slide the clamp clear of the joint. Remove the two knurled socket nuts which secure the exhaust pipe flange at the cylinder head and then pull the pipe forward, out of the port and out of the silencer. The preceding procedure applies equally to the SR500 model. It is suggested, however, that the silencer is removed in addition to the pipe, to prevent accidental damage to the chrome finish. The silencer is supported on two bonded rubber bush mountings.

6   Turn the petrol tap to the 'Off' position (TT and XT models) or the 'On' or 'Res' position (SR models) and detach the petrol feed pipe, and on SR models the additional vacuum pipe. Each pipe is secured at the union by a spring clip, the ears of which should be pinched together to release the tension on the pipe as it is removed. Unscrew the single bolt securing the rear of the petrol tank to the frame. On SR500 models the tank may be eased backwards off the support rubbers at the tank front, and then lifted away. The petrol tank fitted to all other models is further retained by two bolts at the front.

7   Slacken the adjuster screws at the lower end of the two throttle cables. Slip one adjuster out of the cable anchor and detach the nipple from the control pulley. Repeat the operation for the second cable. Slacken fully the carburettor/air filter hose screw clip and remove the two socket screws which retain the inlet stub flange at the cylinder head. Pull the carburettor out at the front so that the inlet stub clears the cylinder head. The carburettor can then be pulled out of the air hose. If difficulty is encountered in removing the carburettor due to the limited space between the air filter box and cylinder head, the air filter box may be removed.

8   Where fitted, the breather chamber and associated hoses should be detached. The hoses are retained by screw clips as is the chamber. Where no chamber is fitted, the breather hose should be detached from the union at the rear of the crankcase.

9   Detach the gearchange lever from the left-hand side of the machine. The gear lever is secured on the splined shaft by a pinch bolt. The bolt must be removed to allow the pedal to be pulled from place. Slacken evenly and remove the screws holding the final drive sprocket cover in place, and lift the cover from position. Similarly, detach the flywheel generator cover. On machines fitted with an endless final drive chain the sprocket and chain must be removed simultaneously. Where a chain with a master link is used it is suggested that the sprocket nut is loosened before the chain is separated because the chain provides a useful means of preventing the sprocket rotating. Bend down the sides of the locknut which secure the final drive sprocket nut and apply a spanner to the nut. To prevent sprocket rotation select top gear – after temporarily refitting the gear lever – and apply the rear brake.

10   Detach the right-hand forward footrest. On SR500 models the footrest is held by two dome nuts, and is supported by two studs. The footrest on all other models is fitted to a splined shaft, clamped in place by a pinch bolt and prevented from slipping off the shaft by a bolt and large washer. Remove the rear brake pedal from the splined pivot shaft after loosening the pinch bolt.Note the spacer behind the pedal boss on the shaft. This is easily mislaid.

11   Apply a suitable size of open ended spanner to the clutch operating arm where it projects from the crankcase. Operate the handlebar lever and then release it, keeping the arm in the disengagement position. The clutch cable may now be displaced from the nipple holder. The security tang, which prevents the cable from jumping out of the holder, may need to be bent down to free the cable. Withdraw the complete cable from the anchor tunnel in the casing.

12   Apply the decompression lever (exhaust lifter) at the handlebars and prevent the operating arm on the cylinder head rocker box from returning by judicious use of a suitable lever. Release the handlebar lever, and making use of the cable slack produced, disconnect the cable from the arm. Where fitted, the tachometer cable should be disconnected from the rocker box. The cable is secured by a spring clip lying in a groove in the cable housing mouth. Displace the clip by pinching together the ears, and then pull the cable out of the tunnel.

13   Detach the oil feed hose from the left-hand side of the engine by removing the single clamp screw and the two socket screws which pass through the junction flange. Similarly, detach the oil return hose from the union at the crankcase. When removing either hose, be prepared for the leakage of a small amount of oil.

14   Trace the wires up from the flywheel generator, contact breaker housing (except SR500 models) and the neutral indicator switch (except TT500 models) and disconnect them at the appropriate snap or block connectors adjacent to the frame downtubes. Pull the suppressor cap from the top of the sparking plug. Detach the narrow bore breather tube from the top of the crankcase by easing out the plastic union.

15 The engine is supported by four mounting bolts. The two upper bolts pass through detachable mounting plates on the right-hand side of the frame. SR500 models are fitted with an additional detachable plate on the left-hand side of the frame, through which the front upper bolt passes. All models also utilise a head steady bracket which interconnects the frame top tube with the rocker cover. The steady bracket is in the form of two separate plates or a single fabricated assembly, depending on the model to which it is fitted. Commence engine removal by detaching the head steady bracket. Remove the nuts from the four engine mounting bolts and detach the removable plates. Withdraw the engine bolts one at a time, lifting the engine as required to prevent tying. If the bolts require drifting out, take care not to damage their threads. Before lifting the engine from position, ensure that all cables, wires and other connections are detached and tucked out of the way so that they do not become snagged. The engine is not particularly heavy, but is difficult to grasp. It is suggested that an assistant is to hand when the engine is lifted out. Remove the engine towards the right-hand side of the machine to take advantage of the extra room provided by the detachable engine plates.

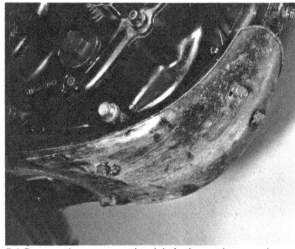

5.1 Remove the sump guard and drain the crankcase and ...

5.2 ... the oil tank of oil

5.3a Disconnect the oil feed pipe and ...

5.3b ...unscrew the combined union/filter

5.4a The dualseat is held by a single bolt at each side

5.4b Release the battery strap and disconnect the battery leads

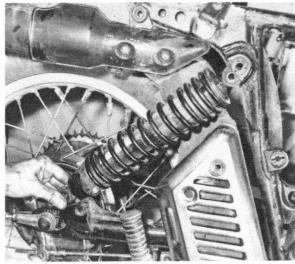

5.5a Unscrew the two right-hand suspension strut bolts

5.5b Remove the exhaust pipe flange knurled screws and the ...

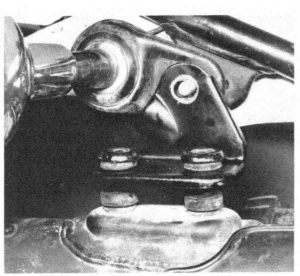

5.5c ... spark arrester support nut

5.5d Remove the upper and ...

5.5e ... lower silencer mounting bolts and ...

5.5f ... lift the complete unit away

5.7a Slacken the cable adjusters and disconnect the cables

5.7b Remove the carburettor after slackening the hose clamps

5.8 Detach the breather chamber from the two unions

5.9 Prevent sprocket rotation by engaging gear and applying brake

5.11 Bend down tang to enable clutch cable removal

5.12a Displace decompressor outer cable and detach nipple

5.12b Prise out the circlip and ...

5.12c ... pull out the tachometer drive cable

6   Dismantling the engine and gearbox: general

1   Before commencing work on the engine unit, the external surfaces should be cleaned thoroughly. A motorcycle engine has very little protection from road grit and other foreign matter, which will find its way into the dismantled engine if this simple precaution is not taken. One of the proprietary cleaning compounds, such as "Gunk" or "Jizer" can be used to good effect, particularly if the compound is permitted to work into the film of oil and grease before it is washed away. Special care is necessary when washing down to prevent water from entering the now exposed parts of the engine unit.

2   Never use undue force to remove any stubborn part unless specific mention is made of this requirement. There is invariably good reason why a part is difficult to remove, often because the dismantling operation has been tackled in the wrong sequence.

3   Before commencing dismantling, make arrangements for storing separately the various sub-assemblies and ancillary components, to prevent confusion on reassembly. Where possible, replace nuts and washers on the studs or bolts from which they were removed and refit nuts, bolts and washers to their components. This too will facilitate straightforward reassembly.

4   Identical sub-assemblies, such as valve springs and collets or rocker arms and pins etc should be stored separately, to prevent accidental transposition and to enable them to be fitted in their original locations.

7   Dismantling the engine/gearbox unit: removing the overhead camshaft and cylinder head

1   Remove the two valve covers. Each is retained by two screws and has a rubber seal on the inside to maintain an oiltight joint. Unscrew the sparking plug and rotate the engine so that both valves are closed fully. This will prevent distortion to the rocker box that would be inevitable if the rocker box retaining nuts were released with one or both valves partially open.

2   Disconnect the rocker box oil feed pipe by removing the two banjo bolts and the washers. Slacken the cylinder head (rocker box) nuts and socket screws evenly, a little at a time, following the sequence given in the accompanying illustration. After removing all the nuts and bolts, separate the rocker box from the mating surface, using a rawhide mallet. If this fails to lift the cover, rotate the engine through about 360°. The rotation of the camshaft will displace the cover. Lift out the cover end plug.

3   Unscrew the cam chain tensioner cover and slacken the locknut. (The largest nut on the assembly). Remove the tensioner unit, noting the relative position of the various components. Apply an open-ended spanner to the large hexagon on the camshaft, and holding the camshaft still, slacken and remove the sprocket retaining centre bolt, together with the piston position indicator (where fitted). If the bolt is very tight, some difficulty may be encountered in restraining the camshaft by means of the hexagon provided. The camshaft will ride up, out of the bearing housings, and this may cause damage to the cam chain. To overcome this problem, temporarily replace the rocker box, securing it with a few equally spaced nuts. Place a spanner on the alternator centre nut to prevent engine rotation and then loosen the sprocket bolt by inserting a spanner through the aperture in the end of the rocker box. Once the bolt has been loosened, remove the rocker box.

4   If a top-end overhaul is envisaged, the cam chain must be prevented from dropping down into the crankcase when the sprocket is removed. To prevent this, secure the chain with a suitable length of stiff wire. Lift the sprocket off the end of the camshaft taking great care – in all cases – that the small drive pin, which is a push fit in the rear of the sprocket and camshaft boss, does not drop into the crankcase. Store the drive pin in a

safe place. Lift the camshaft out of the bearing housings, complete with the camshaft bearings.

5   Slacken and remove the remaining cylinder head securing screws and the domed nut. Using a rawhide mallet, break the joint between the cylinder head and the cylinder barrel. Take care to strike only those portions of the cylinder head which are well supported by lugs and webs or damaged fins may result. Lift the cylinder head up off the holding down studs and away from the engine. To prevent the cam chain from being dropped down the tunnel, slide a screwdriver in between the cylinder head and barrel as soon as there is sufficient room.

7.4a Tether the cam chain when doing a top end overhaul only

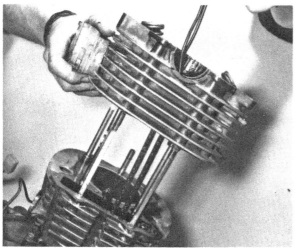

7.4b Chain may be lowered through tunnel when removing head

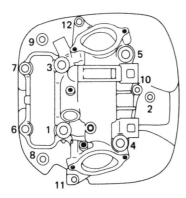

Fig. 1.1 Cylinder head bolt and screw tightening sequence

---

8   Dismantling the engine/gearbox unit: removing the cylinder barrel and piston

1   The cylinder barrel is secured by four sleeve nuts, recessed in the fins and three short socket bolts along the right-hand side of the barrel flange. Slacken the three bolts first and then loosen the four nuts evenly, in a diagonal sequence. Remove the nuts and bolts.

2   Ease the cylinder barrel gently upwards, sliding it along the holding down studs. Take care to support the piston and rings as they emerge from the cylinder bore, otherwise there is risk of damage or ring breakage. If the crankcases are not to be separated, it is advisable to pad the crankcase mouth with clean rags before the piston is withdrawn from the bore, in case the piston rings have broken. This will prevent sections of broken

ring from falling into the crankcase.

3   Prise one of the gudgeon pin circlips out of position, then press the gudgeon pin out of the small end bearing, through the piston boss. If the pin is a tight fit, it may be necessary to warm the piston so that the grip on the gudgeon pin is released. A rag soaked in warm water will suffice, if it is placed on the piston crown. The piston may be lifted from the connecting rod once the gudgeon pin is clear of the small-end eye. Note the arrow mark on the piston crown. On reassembly, the piston must be fitted with the arrow facing forwards.

4   If the gudgeon pin is still a tight fit after warming the piston, it can be lightly tapped out of position with a hammer and soft metal drift. Do NOT use excess force and make sure the connecting rod is supported during this operation, or there is risk of it bending.

---

9   Dismantling the engine/gearbox unit: removing the flywheel generator

1   Remove the nut from the centre of the flywheel generator rotor (alternator rotor SR500 models). To prevent crankshaft rotation during loosening, place a close fitting bar through the small-end eye of the connecting rod, and position two wood blocks of equal size across the crankcase mouth, upon which the bar may bear.

2   The flywheel rotor is a tapered fit on the crankshaft end, located by a Woodruff key, and will require pulling from position. The flywheel centre has an internal left-hand thread — provided to take the Yamaha service tool No. 90890-01189. This tool **must** be used to remove the flywheel. There is insufficient room at the flywheel periphery to enable alternative types of pullers, such as a three or two-legged sprocket extractor, to be fitted. No attempt should be made to lever the flywheel from position. This will only result in damaged components. Screw the flywheel puller into the rotor, so that it is fully home. Lock the crankshaft using the close-fitting bar, and tighten down the puller centre screw. If the flywheel is very tight, do not continue tightening the puller. A smart tap on the puller screw head should bring the rotor from position. After rotor removal, prise the Woodruff key from the tapered shaft end and store it in a safe place to prevent loss.

3   Free the generator lead wire from the cable clamps, where necessary loosening the clamp holding screws. Disconnect the lead running to the neutral warning switch; it is held by a single screw. The generator stator on all but the SR500 models is retained by four screws at the periphery. Remove the screws and lift the stator from position, sliding the wiring grommet

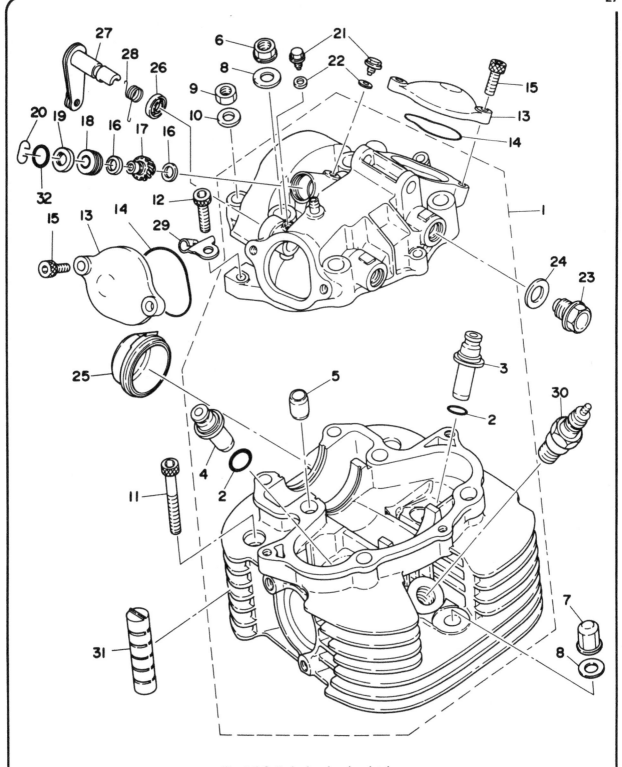

Fig. 1.2 Cylinder head and rocker box

1	Cylinder head assembly	9	Nut – 2 off	17	Tachometer drive gear	25	Piston position indicator cover
2	'O' ring – 2 off	10	Washer – 2 off	18	Bush	26	Oil seal
3	Inlet valve guide	11	Socket screw – 2 off	19	Oil seal	27	Decompressor shaft
4	Exhaust valve guide	12	Socket screw – 3 off	20	Circlip	28	Spring
5	Dowel – 2 off	13	Rocker cover – 2 off	21	Bolt – 2 off	29	Bracket
6	Nut – 4 off	14	'O' ring – 2 off	22	Washer – 2 off	30	Sparking plug
7	Nut	15	Socket screw – 4 off	23	Plug – 2 off	31	Absorber – 2 off
8	Washer – 5 off	16	Washer – 2 off	24	Washer – 2 off	32	'O' ring

from place in the chamber wall. On SR500 models the alternator stator is secured by three screws. The screws pass through elongated holes in the stator plate periphery. The holes allow a limited amount of adjustment to be made to the stator position, which allows alteration of the ignition timing. To prevent loss of ignition timing accuracy, and to aid reassembly, punch mark the stator plate and crankcase before loosening the screws. Take care not to damage the stator coils that are in close proximity to the screws.

10 Dismantling the engine/gearbox unit: removing the contact breaker assembly and ATU – except SR500 models

1   Take off the contact breaker cover, which is retained by two socket screws. Remove the bolt which passes into the contact breaker cam. Mark the contact breaker baseplate in relation to the housing to which it is attached, so that the ignition timing is not altered when reassembly eventually takes place. Then remove the two screws and washers that retain the baseplate in position and lift the complete contact breaker assembly away, together with the flexible lead.

2   Pull the auto-advance mechanism and contact breaker cam from the end of the driveshaft. It is a push fit and is located by means of a pin that keys with the end of the driveshaft. Check the security of the pin in the shaft. If the pin is loose it should be removed and stored in a safe place.

11 Dismantling the engine/gearbox unit: removing the oil filter and gauze sump strainer

1   Tip the engine over so that it leans at an angle to the workbench on the left-hand side. This will prevent any residual oil which has remained in the oil filter chamber from leaking out. Slacken evenly and remove the three cover bolts and lift off the cover. Remove the oil filter element and discard it; the old element should not be reused. Any residual oil may be tipped out into a suitable container.
2   Slacken the screws which hold the sump cover in position. After removal of the screws lift the sump cover away to expose the oil strainer screen. The screen is located in a groove in the cover and a similar groove in the crankcase, and can be pulled from position.

8.2 Pad crankcase mouth with rag before exposing piston rings

8.3 Prise out the gudgeon pin circlips and displace the pin

9.1 Remove the generator nut and washer

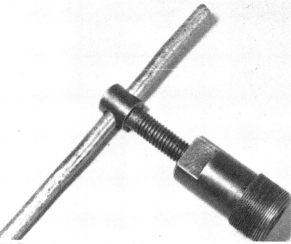
9.2 The flywheel must be withdrawn using this extractor

12 Dismantling the engine/gearbox unit: removing the primary drive cover, primary drive pinion and camshaft drive sprocket

1   Remove the pinch bolt which passes through the kickstart lever boss and pull the lever assembly off the splined shaft. Slacken evenly, in a diagonal sequence, and then remove the primary drive cover screws. Note and mark the position of the screws holding the various cable clips. The primary drive cover may be securely in position owing to the type of gasket used and the two dowels on which it locates. Use a rawhide mallet or the palm of the hand to break the joint. **Do not** insert levers to aid removal, this will only lead to damaged mating surfaces and the resultant oil leaks. When drawing off the cover, push the ATU/contact breaker cam driveshaft inwards, to prevent it coming away with the cover.
2   The cam tensioner blade pivots on a headed pin which is a push fit in the primary drive chamber. Withdraw the pin and lift out the freed blade.
3   The ATU/contact breaker cam driveshaft and the integral driven pinion is a push fit in the casing. Before withdrawing the gear/shaft assembly, note that the driven pinion and the smaller drive pinion with which it meshes are both provided with a punched mark on the outer face of the pinion. These timing marks are provided to enable accurate reassembly, and to ensure that the relative positions of the crankshaft and contact breaker cam are maintained. Neither of these pinions are fitted to the SR500 models.
4   Lock the crankshaft by inserting a close-fitting bar through the small-end eye of the connecting rod, allowing it to bear on two wooden blocks across the crankcase mouth. Remove the crankshaft end nut and washer and then pull off the contact breaker drive pinion (a collar on SR500 models), the primary drive pinion and the timing index plate. Prise out the large Woodruff key from the keyway. The camshaft chain can now be removed and the drive sprocket slid off the shaft.

13 Dismantling the engine/gearbox unit: removing the clutch

1   Slacken the clutch spring bolts evenly, in a diagonal sequence, so that the tension on the pressure plate is reduced in an even manner. Remove the bolts and springs and lift out the pressure plate. Remove the clutch plates either one at a time or as a 'sandwich', noting the alternate sequence of plain and friction plates.
2   Withdraw the mushroom-headed clutch thrust piece from the centre of the clutch shaft. The remaining portion of the clutch operating pushrod may be displaced by tipping the engine over and operating the clutch arm on the crankcase.
3   To enable the clutch centre nut to be loosened, the clutch centre must be prevented from rotating as follows: temporarily refit the final drive sprocket and the gearchange lever, and select top gear. Secure the sprocket by placing a stout lever across one tooth so that the lever engages with a suitably strong lug or boss on the gearbox wall. With the sprocket held in this manner the clutch nut may be loosened. It is recommended that two people carry out this operation; the lever holding the sprocket **must not** be allowed to slip, because damage to the casing will almost inevitably result.
4   After removal of the nut and washer, pull off the clutch centre followed by the heavy thrust washer, the clutch outer drum and the clutch spacer. Fitted behind the clutch are the oil pump drive pinion and kickstart driven pinion. These components may be removed also.

14 Dismantling the engine/gearbox unit: removing the oil pump

1   To enable access to be made to the oil pump securing screws the oil pump driven pinion must first be removed. The pinion is secured by a circlip and a washer. After removing the pinion, note and displace the remaining washer on the shaft.
2   Slacken evenly the three screws which secure the oil pump to the casing. The oil pump has two sets of rotors, and is in fact two separate oil pumps driven by the same shaft. The outer rotors (delivery pump) are housed within a detachable casing which serves as a cover for the inner rotors (scavenge pump) which work in a recess bored directly into the crankcase. After removing the screws pull the oil pump away from the casing. The complete delivery pump and the scavenge pump inner rotor will come away as a unit, leaving the scavenge pump outer rotor in the crankcase. This may be displaced and refitted on the inner rotor. Store the oil pump in a safe place until examination or reassembly is required, and ensure that no foreign matter is allowed to enter the pump.

15 Dismantling the engine/gearbox unit: removing the kick-start assembly

1   Free the kickstart idler pinion by displacing the circlip and washer and then pull it off the shaft.
2   Remove the washer and circlip from the kickstart shaft. Grasp the outer end of the return spring with a pair of pliers and pull the hook off the anchor lug in the casing. Allow the spring to unwind in an anti-clockwise direction in a controlled manner. Withdraw the spring guide from the centre of the spring and displace the inner turned end of the spring from the radial hole in the shaft, so that the spring can be removed.
3   The kickstart shaft, together with the components remaining on it, may now be removed from the casing as a complete unit. Further dismantling for examination or renewal is described in Section 28 of this Chapter.

16 Dismantling the engine/gearbox unit: removing the gearchange mechanism

1   Before the gearchange shaft can be removed from the engine, the 'E' clip, which secures it on the left-hand side of the unit, must be displaced and the washer removed. Grasp the gearchange arm at the right-hand end and pull it out of the casing, together with the centraliser spring.
2   The gearchange quadrant and the pawl arm mounted on it may be removed as a unit, after displacing the 'E' clip. Slide the quadrant off the end of the pivot spindle, simultaneously depressing the pawl arm so that it clears the change pins in the end of the change drum.
3   Tip the engine over onto its right-hand side and unscrew the change drum detent housing from the base of the gearbox. Lift out the detent spring and plunger.

17 Dismantling the engine/gearbox unit: separating the crankcase halves

1   The crankcase halves are secured together by screws passing through from the left-hand side of the engine. Slacken the screws evenly, in a diagonal sequence, to help prevent distortion, then remove the screws. Note and mark the positions of any cable clamps to aid reassembly.
2   Initial separation of the crankcases should be made using a rawhide mallet to break the gasket compound joint. Screwdrivers or other levers **must not** be used to aid separation; this sort of treatment will damage the mating surfaces and result in oil leakage. The cases should be separated so that the right-hand case is lifted away, leaving the crankshaft and gearbox components in the left-hand half. To accomplish this, the right-hand end of the crankshaft should be tapped so that the mainshaft boss leaves the main bearing. The gearbox shaft should be tapped as necessary, as separation progresses. If any difficulty is encountered in separating the crankcase halves check that no casing screws have been left in inadvertently.

18 Dismantling the engine/gearbox unit: removing the gearbox components and the crankshaft

1   Lift the front selector fork rod so that the lower end leaves the recess in the casing, and manoeuvre the rod and selector fork from position. Remove the rear selector fork rod and the two selector forks in a similar manner. The gearchange drum is now free to be pulled out of the bearing housing.
2   Grasp both gearshaft clusters and lift the two assemblies out simultaneously, still meshed with one another. Make a careful note of any endfloat shims fitted to either end of the shafts. The shims must be fitted in the original positions on subsequent reassembly.
3   The crankshaft left-hand mainshaft is a light drive fit in the main bearing, consequently the crankshaft must be pressed or tapped from position using a rawhide mallet. Care must be taken to support the connecting rod and also to prevent damage to the mainshaft thread. This is particularly important where the thread is external. If necessary refit the alternator nut so that the nut outer face is flush with the end of the shaft.

19 Examination and renovation: general

1   Before examining the parts of the dismantled engine unit for wear, it is essential that they should be cleaned thoroughly. Use a paraffin/petrol mix to remove all traces of old oil and sludge that may have accumulated within the engine.
2   Examine the crankcase castings for cracks or other signs of damage. If a crack is discovered, it will require professional repair.
3   Examine carefully each part to determine the extent of wear, if necessary checking with the tolerance figures listed in the Specifications Section of this Chapter, or accompanying the text.
4   Use a clean, lint-free rag for cleaning and drying the various components, otherwise there is risk of small particles obstructing the internal oilways.

20 Examination and renovation: big-end and main bearings

1   Failure of the big-end bearing is invariably accompanied by a knock from within the crankcase that progressively becomes worse. Some vibration will also be experienced. To assess overall wear of the big-end bearings measure the deflection of the connecting rod at one face of the small-end eye, in-line with the crankshaft. If the movement exceeds 2.0 mm (0.08 in) the bearing is in need of renewal. Check also the endfloat of the big-end eye between the flywheels, using a feeler gauge. If axial play is greater than 0.7 mm (0.028 in) Yamaha recommend that the worn components are replaced. Do not run the machine with a worn big-end bearing, otherwise there is risk of breaking the connecting rod or crankshaft.
2   It is not possible to separate the flywheel assembly in order to replace the bearing because the parallel sided crankpin is pressed into the flywheels. Big-end repair should be entrusted to a Yamaha Service Agent, who will have the necessary repair or replacement facilities.
3   Failure of the main bearings is usually evident in the form of an audible rumble from the bottom end of the engine, accompanied by vibration. The vibration will be most noticeable through the footrests.
4   The crankshaft main bearings are of the journal ball type. If wear is evident in the form of play or if the bearings feel rough as they are rotated, replacement is necessary.
5   Both main bearings are a tight drive fit in their bearing housings, the right-hand bearing being secured by a plate held by two screws. Each bearing may be drifted from the casing, using a suitable thick-walled tubular drift. To aid bearing removal the crankcase half being attended to should be heated in an oven to 95°-125°C (200°-250°F) **Do not** use a blow

lamp or other means of inducing localised heat, because this may distort the casting.
6   The right-hand bearing is fitted with an oil seal on the outside. This should be prised from position before bearing removal, and discarded.
7   New main bearings may be fitted by reversing the dismantling procedure, again heating the cases first. The bearings should be driven inwards with the manufacturer's marks or numbers facing outwards ie towards the drift. Ensure that the bearing remains square as it is being driven home, otherwise damage to the housings will result. In practice, it will be found – provided that the cases are at the correct heat – that the bearing is a very light drive fit.

21 Examination and renovation: cylinder barrel

1   The usual indications of a badly worn cylinder barrel and piston are excessive oil consumption and piston slap, a metallic rattle that occurs when there is little or no load on the engine. If the top of the bore of the cylinder barrel is examined carefully, it will be found that there is a ridge on the thrust side, the depth of which will vary according to the amount of wear that has taken place. This marks the limit of travel of the uppermost piston ring.
2   Measure the bore diameter just below the ridge, in line with the gudgeon pin axis, using an internal micrometer. Take a second measurement at right angles to the first measurement. Take two further sets of readings half-way down the bore and about 25 mm (1 in) from the lower end of the skirt. If any reading exceeds the specified wear limit or the difference between two readings taken at different levels exceed the stated amount a rebore is required.
3   If an internal micrometer is not available, the amount of cylinder bore wear can be measured by inserting the piston without rings so that it is approximately $\frac{3}{4}$ inch from the top of the bore. If it is possible to insert a 0.10 mm (0.004 in) feeler gauge between the piston and the cylinder wall on the thrust side of the piston, remedial action must be taken.
4   Check the surface of the cylinder bore for score marks or any other damage that may have resulted from an earlier engine seizure or displacement of the gudgeon pin. A rebore will be necessary to remove any deep indentations, irrespective of the amount of bore wear, otherwise a compression leak will occur.
5   Check the external cooling fins are not clogged with oil or road dirt; otherwise the engine will overheat.

22 Examination and renovation: piston and piston rings

1   If a rebore is necessary, the existing piston and piston rings can be disregarded because they will be replaced with their oversize equivalents as a matter of course.
2   Remove all traces of carbon from the piston crown, using a soft scraper to ensure the surface is not marked. Finish off by polishing the crown with metal polish so that carbon does not adhere so easily in the future. Never use emery cloth.
3   Piston wear usually occurs at the skirt or lower end of the piston and takes the form of vertical streaks or score marks on the thrust side. There may also be some variation in the thickness of the skirt.
4   The piston ring grooves may also become enlarged in use, allowing the piston rings to have greater side float. If the clearance exceeds that given in the Specifications for each ring, the piston is due for replacement. It is unusual for this amount of wear to occur on its own.
5   Piston ring wear is measured by removing the rings from the piston and inserting them in the cylinder bore using the crown of the piston to locate them approximately 40 mm (1$\frac{1}{2}$ inches) from the top of the bore. Make sure they rest square with the bore. Measure the end gap with a feeler gauge; if the gap exceeds the stated maximum they require replacement,

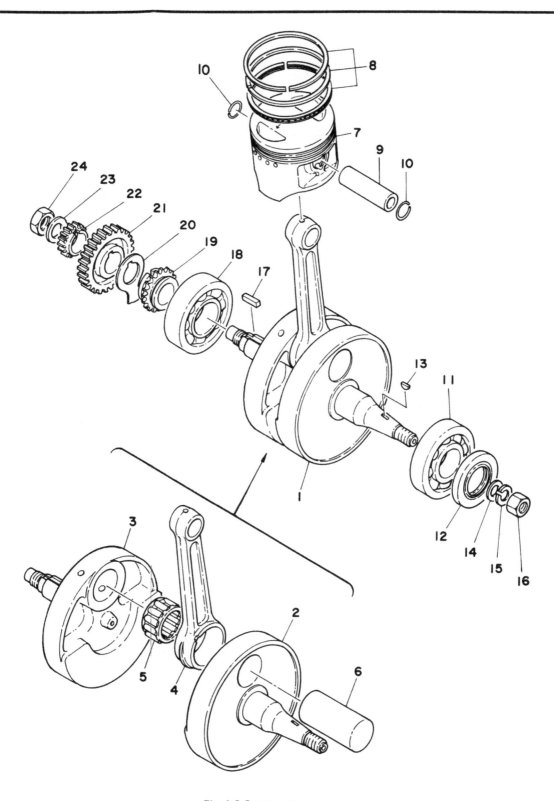

Fig. 1.3 Crankshaft and piston

1	Crankshaft assembly	7	Piston	13	Woodruff key	19	Cam chain drive sprocket
2	Flywheel	8	Piston ring set	14	Washer	20	Timing index plate
3	Flywheel	9	Gudgeon pin	15	Spring washer	21	Primary drive pinion
4	Connecting rod	10	Circlip – 2 off	16	Nut	22	Contact breaker drive pinion
5	Bearing	11	Main bearing	17	Woodruff key	23	Washer
6	Crankpin	12	Oil seal	18	Main bearing	24	Nut

assuming the cylinder barrel is not in need of a rebore. Due to the relatively low cost involved, and the important function of the piston rings it is considered good practice to renew them as a matter of course when the engine is dismantled.

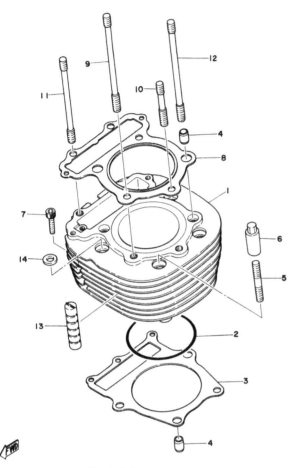

Fig. 1.4 Cylinder barrel

1	Cylinder barrel	8	Cylinder head gasket
2	Cylinder base 'O' ring	9	Stud – 2 off
3	Cylinder base gasket	10	Stud
4	Dowel – 4 off	11	Stud – 2 off
5	Stud – 4 off	12	Stud – 2 off
6	Nut – 4 off	13	Absorber – 2 off
7	Socket screw – 3 off	14	Washer – 4 off

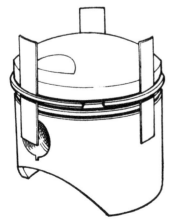

Fig. 1.5 Method of removing and replacing piston rings

23 Examination and renovation: cylinder head and valves

1   It is best to remove all carbon deposits from the combustion chambers before removing the valves for inspection and grinding-in. Use a blunt end chisel or scraper so that the surfaces are not damaged. Finish off with a metal polish to achieve a smooth, shining surface. If a mirror finish is required, a high speed felt mop and polishing soap may be used. A chuck attached to a flexible drive will facilitate the polishing operation.

2   A valve spring compression tool must be used to compress each set of valve springs in turn, thereby allowing the split collets to be removed from the valve cap and the valve springs and caps to be freed. Keep each set of parts separate to avoid the accidental interchange of individual components. There is no danger of inadvertently transposing the valves when refitting them because the valve heads are of different diameters.

3   Before giving the valve and valve seats further attention, check the clearance between each valve stem and the guide in which it operates. The valve stem/guide clearance can be measured with the use of a dial gauge and a new valve. Place the new valve into the guide and measure the amount of shake with the dial gauge tip resting against the top of the stem. If the amount of wear is greater than the wear limit, the guide must be renewed. Measure the valve stem at the point of greatest wear and then measure again at right-angles to the first measurement. If the valve stem is below the service limit it must be renewed.

4   To remove the old valve guide, place the cylinder head in an oven and heat it to about 100°C (212°F). The old guide can now be tapped out from the cylinder side. To prevent distortion of the large alloy casting it is essential that the cylinder head is heated evenly. For this reason an oven **must** be used in preference to a blow torch or other methods of heating. If inexperienced in this type of work, the advice of a Yamaha service Agent should be sought. Before drifting a guide from place, remove any carbon deposits, which may have built up on the guide end projecting into the port. Carbon deposits will impede the progress of the guide and may damage the cylinder head. If possible, use a double diameter drift. The smaller diameter should be close to that of the valve stem, and the larger diameter slightly smaller than that of valve guide. Provided that care is exercised, a parallel shanked drift may be used as a substitute. A new guide may be fitted by reversing the dismantling procedure. Ensure that the 'O' ring is fitted to the guide below the shouldered portion.

5   Valve grinding is a simple task. Commence by smearing a trace of fine valve grinding compound (carborundum paste) on the valve seat and apply a suction tool to the head of the valve. Oil the valve stem and insert the valve in the guide so that the two surfaces to be ground-in make contact with one another. With a semi-rotary motion, grind in the valve head to the seat, using a backward and forward action. Lift the valve occasionally so that the grinding compound is distributed evenly. Repeat the application until an unbroken ring of light grey matt finish is obtained on both valve and seat. This denotes the grinding operation is now complete. Before passing to the next valve, make sure that all traces of the valve grinding compound have been removed from both the valve and its seat and that none has entered the valve guide. If this precaution is not observed, rapid wear will take place due to the highly abrasive nature of the carborundum base.

6   If, after grinding, it is found that the width of the grey seating ring is greater than 2.0 mm (0.08 in) the valve seat must be recut using a special cutting tool. A 30° seat cutter is required to restore the seat width followed by a 45° seat cutter in order to restore the seat width to 1.3 mm (0.05 in) and the seat angle to 45°. Because of the expense of purchasing the two seat cutters and because of the accuracy with which cutting must be carried out, it is strongly recommended that the cylinder head be returned to a Yamaha Service Agent for attention.

7   Where deep pitting of the seat and valve is encountered or

when a new guide has been fitted, the seat should be recut as previously described. The valve face may be ground back on a special grinding machine to an angle of 45°, provided that after grinding, the depth of the valve periphery has not been reduced excessively.

8   Examine the condition of the valve collets and the groove on the valve stem in which they seat. If there is any sign of damage, new parts should be fitted. Check that the valve spring collar is not cracked. If the collets work loose or the collar splits whilst the engine is running, a valve could drop into the cylinder and cause extensive damage.

9   Check the free length of each of the valve springs. The springs have reached their serviceable limit when they have compressed to the limit readings given in the Specifications Section of this Chapter.

10 Reassemble the valve and valve springs by reversing the dismantling procedure. Enure that all the springs are fitted with the close coils downwards towards the cylinder head. Fit new oil seals to each valve guide and oil both the valve stem and the valve guide, prior to reassembly. Take special care to ensure the valve guide oil seal is not damaged when the valve is inserted. As a final check after assembly, give the end of each valve stem a light tap with a hammer, to make sure the split collets have located correctly.

11 Check the cylinder head for straightness, especially if it has shown a tendency to leak oil at the cylinder head joint. If there is any evidence of warpage, provided it is not too great, the cylinder head must be machined flat or a new head will have to be fitted. Most cases of cylinder head warpage can be traced to unequal tensioning of the cylinder head nuts and bolts and by tightening them in incorrect sequence.

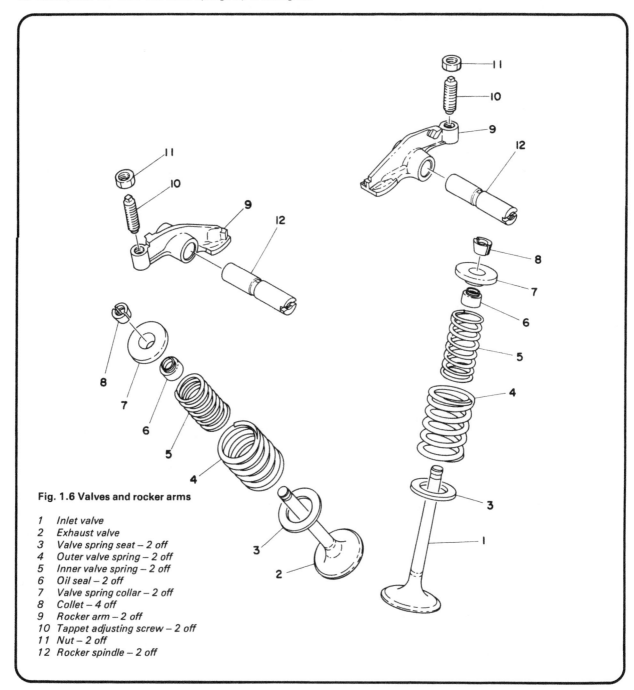

Fig. 1.6 Valves and rocker arms

1   Inlet valve
2   Exhaust valve
3   Valve spring seat – 2 off
4   Outer valve spring – 2 off
5   Inner valve spring – 2 off
6   Oil seal – 2 off
7   Valve spring collar – 2 off
8   Collet – 4 off
9   Rocker arm – 2 off
10 Tappet adjusting screw – 2 off
11 Nut – 2 off
12 Rocker spindle – 2 off

23.10a Fit a new oil seal to the valve guide and ...

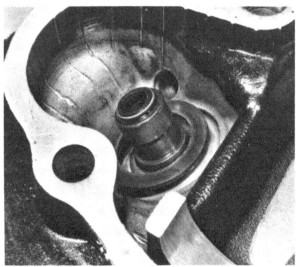

23.10b ... place the spring seating washer in place

23.10c Lubricate the valve stem before insertion in the guide

23.10d Fit both springs with the close pitched coils downwards

23.10e Replace the valve spring collar and ...

23.10f ... use a compressor to allow re-installation of the collets

24 Examination and renovation: overhead camshaft, camshaft sprockets and chain

1 The cams should have a smooth surface and be entirely free from scuff marks or indentations. It is unlikely that severe wear will be encountered during the normal service life of the machine unless the lubrication system has failed, causing the case hardened surface to wear through. If necessary, check with the Specifications given at the beginning of this Chapter and measure the cam height in each case. If either cam is below the service limit, the camshaft must be renewed.

2 Clean both camshaft bearings thoroughly in petrol and then check them for signs of roughness and play or pitting of the balls and tracks. If any of these conditions are in evidence the bearings should be renewed. Both bearings are a tight push fit on the camshaft and bosses and may be pulled or driven from position.

3 Make sure the timing marks on the upper camshaft sprocket are clearly visible, because they are easily obscured by old oil. It will be necessary to refer to these marks during engine reassembly. Examine the camshaft chain sprockets for worn, broken or chipped teeth, an unusual occurrence that can often be attributed to the presence of foreign bodies or particles from some other broken engine component.

4 Examine the camshaft chain for excessive wear or cracked or broken rollers. An indication of wear is given by the extent to which the chain can be bent sideways; if a pronounced curve is evident, the chain should be renewed.

5 The chain tensioner, like the chain, does not normally give trouble because it is well lubricated. Check that the coating of the tensioner strip has not worn through and that the tensioner rod slides freely in the tensioner body. Examine also the chain guide for wear. In cases of doubt, err on the side of safety and renew without question. Although a rare happening, a broken camshaft chain may cause extensive damage, with risk of the valve entangling with the piston.

25 Examination and renovation: rocker arms and spindles

1 It is unlikely that excessive wear will occur in either the rocker arms or the rocker shafts unless the flow of oil has been impeded or the machine has covered a very large mileage. A clicking noise from the rocker area is the usual symptom of wear in the rocker gear, which should not be confused with a somewhat similar noise caused by excessive valve clearances.

2 If any shake is present and the rocker arm is loose on its shaft, a new rocker arm and/or shaft should be fitted.

3 Check the tip of each rocker arm at the point where the arm makes contact with the cam. If signs of cracking, scuffing or breakthrough in the case hardened surface are evident, fit a new replacement.

4 Check also the thread of the tappet adjusting screw, the thread of the rocker arm into which it fits and the thread of the locknut. The hardened end of the tappet adjuster must also be in good condition.

5 If required, the rocker spindles may be removed and the rocker arms detached. Each spindle is secured by a hexagonal plug which screws into the rocker box. The end of each spindle is threaded to enable a 6 mm casing screw to be inserted and so aid withdrawal of the spindles. When refitting a spindle, ensure that the milled slot in the left-hand end is positioned so that it runs horizontally.

6 The decompression facility for easy starting is controlled by a cam lever, which, when operated, bears against the exhaust rocker arm and so prevents the valve from closing fully. If the lever has been damaged or has worn to the extent where it no longer functions correctly, it should be removed from the rocker box and renewed. The shaft is secured by a location bolt passing into the rocker box. After removal of the bolt the shaft may be withdrawn.

25.5a Each rocker spindle is secured by a plug

25.5b Spindle is threaded to take screw for easy extraction

25.5c Check each end of the rocker arm for wear

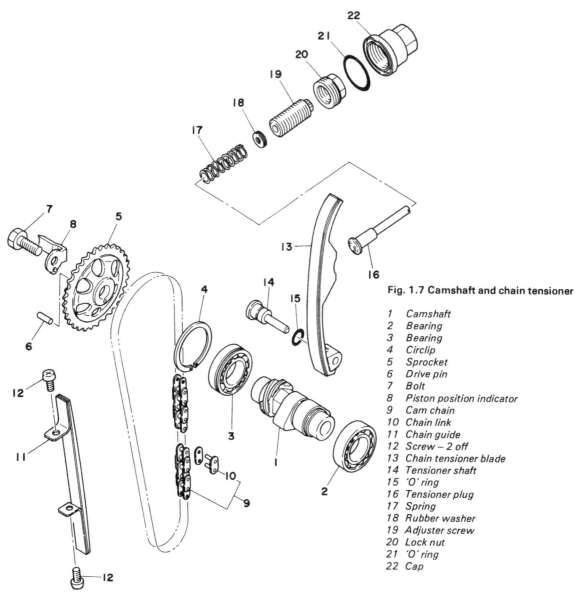

Fig. 1.7 Camshaft and chain tensioner

1   Camshaft
2   Bearing
3   Bearing
4   Circlip
5   Sprocket
6   Drive pin
7   Bolt
8   Piston position indicator
9   Cam chain
10 Chain link
11 Chain guide
12 Screw – 2 off
13 Chain tensioner blade
14 Tensioner shaft
15 'O' ring
16 Tensioner plug
17 Spring
18 Rubber washer
19 Adjuster screw
20 Lock nut
21 'O' ring
22 Cap

26 Examination and renovation: tachometer drive assembly

1   The worm drive to the tachometer is an integral part of the exhaust camshaft which meshes with a pinion attached to the cylinder head cover. If the worm is damaged or badly worn, it will be necessary to renew the camshaft complete.

2   The driveshaft and pinion are a single part retained in the cylinder head in a bush housing which is secured by an internal circlip. Renewal is therefore straightforward. It is unlikely that wear will develop on either the drive or driven pinion as both are well lubricated and lightly loaded.

26.2 Tachometer drive shaft is secured in tunnel by a bolt

27 Examination and renovation: gearbox components

1 It should not be necessary to dismantle either of the gear clusters unless damage has occurred to any of the pinions or if the caged needle roller bearings require attention.

2 The accompanying illustration shows how both clusters of the gearbox are assembled on their respective shafts. It is imperative that the gear clusters, including the thrust washers, are assembled in EXACTLY the correct sequence, otherwise constant gear selection problems will occur. In order to eliminate the risk of misplacement, make rough sketches as the clusters are dismantled. Also strip and rebuild as soon as possible to reduce any confusion which might occur at a later date.

3 When dismantling the gear shafts, the journal ball bearings may be pulled from position, using a standard two or three-legged sprocket puller. The gear pinions on each shaft may be removed in sequence, displacing as necessary the washers and retaining circlips. Refer to the accompanying sequence of photographs.

4 Give the gearbox components a close visual inspection for signs of wear or damage such as broken or chipped teeth, worn dogs, damaged or worn splines and bent selectors. Replace any parts found unserviceable because they cannot be reclaimed in a satisfactory manner.

5 The gearbox bearings must be free from play and show no signs of roughness when they are rotated. After thorough washing in petrol the bearings should be examined for roughness and play. Also check for pitting on the roller tracks.

6 It is advisable to renew the gearbox oil seals irrespective of their condition. Should a re-used seal fail at a later date, a considerable amount of work is involved to gain access to renew it.

7 Check the gear selector rods for straightness by rolling them on a sheet of plate glass. A bent rod will cause difficulty in selecting gears and will make the gear change particularly heavy.

8 The selector forks should be examined closely, to ensure that they are not bent or badly worn. The pegs which engage with the cam channels are not detachable and when they are worn the forks must be renewed. Under normal conditions, the gear selector mechanism is unlikely to wear quickly, unless the gearbox oil level has been allowed to become low.

9 The tracks in the selector drum, with which the selector forks engage, should not show any undue signs of wear unless neglect has led to under lubrication of the gearbox. Check the tension of the gearchange pawl, gearchange lever, centraliser and drum stopper detent springs. Weakness in the springs will lead to imprecise gear selection.

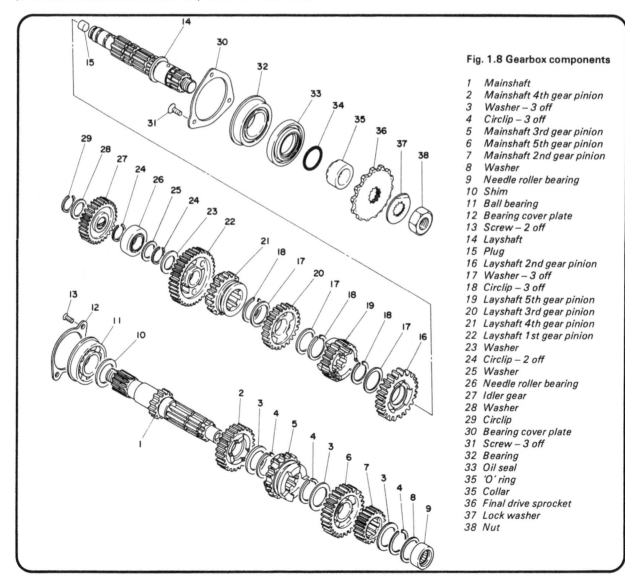

Fig. 1.8 Gearbox components

1 Mainshaft
2 Mainshaft 4th gear pinion
3 Washer – 3 off
4 Circlip – 3 off
5 Mainshaft 3rd gear pinion
6 Mainshaft 5th gear pinion
7 Mainshaft 2nd gear pinion
8 Washer
9 Needle roller bearing
10 Shim
11 Ball bearing
12 Bearing cover plate
13 Screw – 2 off
14 Layshaft
15 Plug
16 Layshaft 2nd gear pinion
17 Washer – 3 off
18 Circlip – 3 off
19 Layshaft 5th gear pinion
20 Layshaft 3rd gear pinion
21 Layshaft 4th gear pinion
22 Layshaft 1st gear pinion
23 Washer
24 Circlip – 2 off
25 Washer
26 Needle roller bearing
27 Idler gear
28 Washer
29 Circlip
30 Bearing cover plate
31 Screw – 3 off
32 Bearing
33 Oil seal
35 'O' ring
35 Collar
36 Final drive sprocket
37 Lock washer
38 Nut

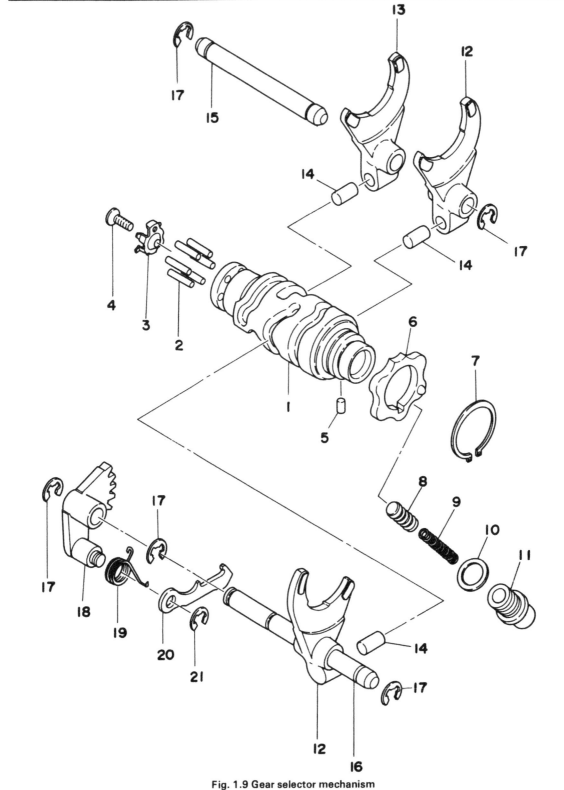

Fig. 1.9 Gear selector mechanism

1	Gearchange drum	8	Detent plunger	15	Selector fork rod
2	Pin – 6 off	9	Spring	16	Selector fork rod
3	Pin retainer plate	10	Sealing washer	17	'E' clip – 5 off
4	Screw	11	Detent plunger housing	18	Selector quadrant
5	Pin	12	Selector fork – 2 off	19	Pawl spring
6	Drum stopper plate	13	Selector fork	20	Change pawl
7	Circlip	14	Guide pin – 3 off	21	'E' clip

27.2 Complete mainshaft and layshaft assemblies

27.3a Remove the washer from clutch end of the mainshaft

27.3b Displace the washer and circlip and ...

27.3c ... pull off the 2nd gear pinion and ...

27.3d ... the 5th gear pinion

27.3e Remove the washer and prise off the circlip to ...

27.3f ... free the 3rd gear pinion

27.3g The 4th gear pinion is held by a circlip and washer

27.3h Remove the circlip and washers from the end of the layshaft ...

27.3i ... withdraw the first gear pinion

27.3j Remove the 4th gear pinion, circlip and washer to free ...

27.3k ... the 3rd gear pinion

27.3l Remaining pinions are the 5th gear pinion and the larger 2nd gear pinion

28 Examination and renovation: kickstart components

1 Check the condition of the kickstart components. If slipping has been encountered a worn ratchet and pawl will invariably be traced as the cause. Any other damage or wear to the components will be self-evident. If either the ratchet or pawl is found to be faulty, both components must be replaced as a pair. Examine the kickstart return spring, which should be renewed if there is any doubt about its condition.
2 The kickstart shaft assembly may be dismantled into its individual components as follows: displace the thrust washer and circlip from the inner end of the shaft and slide off the second washer, pawl engagement spring and the pawl piece. From the opposite end of the shaft remove the return spring guard plate, circlip, shim and kickstart ratchet pinion.
3 The components may be refitted by reversing the dismantling sequence. Note that when refitting the pawl piece to the shaft, the punch mark on each should be in alignment with each other. If this alignment is not correct, the kickstart will not operate correctly.
4 The pawl guide plate and kickstart stop plate, which are still affixed to the primary drive chamber wall, should be inspected for damage or wear. The two bolts passing through both plates are secured by a double locking plate, the ears of which must be bent down away from the bolt heads, before attempting to loosen the bolts.
5 Check the condition of the kickstart idler pinion and the kickstart driven pinion. If there is evidence of excessive wear or broken teeth, the gear in question should be renewed.

29 Examination and renovation: clutch assembly

1 After an extended period of use the clutch linings will wear and promote clutch slip. The clutch plates should be measured with a vernier gauge or pair of calipers to ascertain the extent of wear. The measurement of the thickness for the inserted (friction) plates and the maximum wear limits are given in the Specifications section of this Chapter. If the plate width is less than the specified minimum, then the plate must be renewed.
2 The plain clutch plates should not show any evidence of overheating (blueing). If they do, check them for overall flatness by placing each plate on a flat surface. In the event of the plates being warped by more than 0.05 mm (0.002 in) they should be renewed.
3 Check the free length of each clutch spring. If the springs have shortened (set) to a length less than the specified minimum, they must be renewed, preferably as a set.
4 Examine the amount of play between the clutch spacer, and the clutch shaft and outer drum. Wear of the spacer may be the cause of clutch noise.
5 Check the condition of the slots in the outer surface of the clutch centre and the inner surfaces of the outer drum. In an extreme case, clutch chatter may have caused the tongues of the inserted plates to make indentations in the slots of the outer drum, or the tongues of the plain plates to indent the slots of the clutch centre. These indentations will trap the clutch plates as they are freed and impair clutch action. If the damage is only slight the indentations can be removed by careful work with a file and the burrs removed from the tongues of the clutch plates in similar fashion. More extensive damage will necessitate renewal of the parts concerned.
6 The clutch release mechanism in the crankcase does not normally require attention. After extended mileage, however, developed indentation of the lifting shaft cam face may impair clutch operation or pushrod adjustment. To remove the shaft, slacken the adjuster nut and unscrew the adjuster screw fully. The shaft can then be drawn out.

28.2a Displace the circlip to free the washer spring and pawl

28.2b Kickstart wheel is held by a circlip and washer

28.3 Punch mark on pawl must align with mark on shaft shoulder

30 Examination and renovation: crankcase covers

1   The right-hand and left-hand crankcase covers and the inspection covers are unlikely to become damaged unless the machine is dropped or involved in an accident. Cracks in a casing can be repaired easily by special aluminium welding, provided the damage is not too extensive and care is taken to prevent distortion.

2   On SR500 models, the covers are lightly polished and lacquered before leaving the factory. Badly scratched covers can be refurbished using a single cut file treated with chalk to prevent clogging, and finished off with fine emery paper and metal polish or aluminium cleaner. If required, the cases can be relacquered, using an aerosol paint spray. All other models are fitted with engines finished in a light covering of matt black heat resistant paint. Damaged areas are easily refurbished, using a proprietary matt paint contained in an aerosol can.

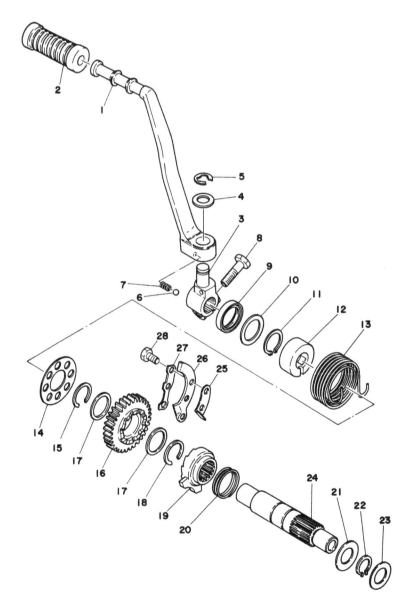

Fig. 1.10 Kickstart mechanism

1   Kickstart lever
2   Rubber
3   Kickstart lever boss
4   Washer
5   'E' clip
6   Steel ball
7   Spring
8   Pinch bolt
9   Oil seal
10   Plain washer
11   Circlip
12   Spring guide
13   Return spring
14   Cover
15   Circlip
16   Kickstart pinion
17   Washer – 2 off
18   Circlip
19   Ratchet wheel
20   Spring
21   Washer
22   Circlip
23   Washer
24   Kickstart shaft
25   Kickstart stop plate
26   Pawl guide plate
27   Locking plate
28   Bolt – 3 off

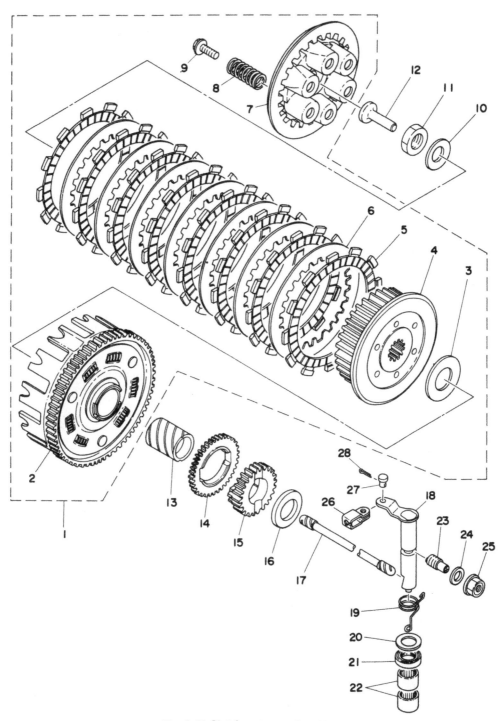

Fig. 1.11 Clutch component parts

1	Clutch assembly	
2	Clutch outer drum	
3	Thrust plate	
4	Clutch boss	
5	Friction plate – 8 off	
6	Plain plate – 7 off	
7	Pressure plate	
8	Spring – 6 off	
9	Screw and washer – 6 off	
10	Beleville washer	

11	Nut
12	Pushrod
13	Spacer
14	Oil pump drive pinion
15	Kickstart driven pinion
16	Thrust plate
17	Pushrod
18	Clutch operating shaft
19	Spring

20	Washer
21	Oil seal
22	Bearing – 2 off
23	Adjusting screw
24	Washer
25	Nut
26	Link arm
27	Pin
28	Split pin

31 Engine reassembly: general

1   Before reassembly of the engine/gearbox unit is commenced, the various component parts should be cleaned thoroughly and placed on a sheet of clean paper, close to the working area.

2   Make sure all traces of old gaskets have been removed and that the mating surfaces are clean and undamaged. One of the best ways to remove old gasket cement is to apply a rag soaked in methylated spirit. This acts as a solvent and will ensure that the cement is removed without resorting to scraping and the consequent risk of damage. If a gasket becomes bonded to the surface through the effects of heat and age, a new sharp scalpel blade should be used to effect removal. Old gasket compound can also be removed using a soft brass wire brush of the type used for cleaning suede shoes. A considerable amount of scrubbing can take place without fear of damaging the mating surfaces.

3   Gather together all the necessary tools and have available an oil can filled with clean engine oil. Make sure that all new gaskets and oil seals are to hand, also all replacement parts required. Nothing is more frustrating than having to stop in the middle of a reassembly sequence because a vital gasket or replacement has been overlooked.

4   Make sure that the reassembly area is clean and that there is adequate working space. Refer to the torque and clearance settings wherever they are given. Many of the smaller bolts are easily sheared if overtightened. Always use the correct size screwdriver or bit for the crosshead screws never an ordinary screwdriver or punch. If the existing screws show evidence of maltreatment in the past, it is advisable to renew them as a complete set.

32 Reassembling the engine/gearbox unit: replacing the crankshaft and gearbox components

1   Place the left-hand crankcase half on the workbench, supported on blocks so that when fitted, the crankshaft will not be obstructed by the workbench surface. Check that the bearings within the casing are correctly positioned and lubricate them with clean engine oil.

2   Lift up the crankshaft and insert the left-hand mainshaft through the main bearing. Because the mainshaft boss is a tight fit in the bearing inner race, the crankshaft must be driven into position, using a rawhide mallet. Strike the right-hand mainshaft, ensuring that the crankshaft as a whole is kept

square with the bearing and that the connecting rod does not foul the edge of the crankcase mating surface.

3   Before being refitted into the crankcase half, the two gearshaft assemblies must be assembled as complete units. Check that the 'O' ring is fitted to the groove provided close to the end of the layshaft.

4   Place the gearshaft end-float shims (where fitted) into the casing so that they lie on the inner races of their respective bearings. The mainshaft and layshaft should be meshed together and fitted into the casing simultaneously. Ensure that the shafts locate with any shims.

5   If the detent quadrant was removed from the gearchange drum, it should now be refitted. The quadrant should be positioned so that the recess in the bore locates with the drive pin projecting from the change drum boss. Note that the 'pip' projecting from the detent quadrant face operates the neutral indicator switch and must therefore be positioned towards the outside. Install the gearchange drum in the casing and then refit the front selector rod which carries one selector fork, and the rear selector rod which carries two selector forks. The forks should be engaged with the drum and the gears before the rods are pushed home into the recesses provided in the gearbox wall.

32.1 Position the left-hand casing on blocks on the bench

32.2 Install the crankshaft in the casing

32.3 Do not omit any gearshaft end-float shims

32.4 Fit mainshaft and layshaft assemblies simultaneously

32.5a Position circlip on gearchange detent quadrant as shown

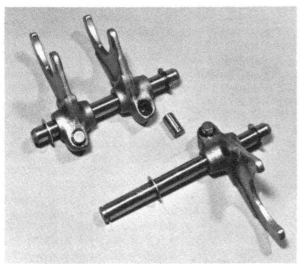

32.5b Relative positions of selector forks and rods are as shown

32.5c Install the gearchange drum

32.5d Locate the two forks and rod and ...

32.5e ... fit the single fork and rod

33 Reassembling the engine/gearbox unit: joining the crankcase halves

1 Check that both crankcase half mating surfaces are absolutely clean and free of old jointing compound. Fit the two locating dowels to the holes provided in either half and apply a coat of gasket compound to the other mating surface. Place a new 'O' ring in the recess in the right-hand casing.

2 Lubricate thoroughly the gearbox components, the crankshaft and all the bearings with clean engine oil. Position the right-hand crankcase half over the partially assembled engine so that the bearings and shafts align and then lower it into position on the shafts. Using a rawhide mallet tap the casing down into place, keeping the two crankcase halves square with one another. Rotate the various shafts during this operation, to facilitate assembly.

3 When the cases have seated, insert the casing screws together with any cable clips that were removed during dismantling. Tighten the screws in a diagonal sequence, to avoid distortion. When the screws have been tightened fully check that all the shafts will still rotate freely.

34 Reassembling the engine/gearbox unit: replacing the flywheel generator

1 Position the generator stator over the crankshaft end so that it rests in the stator chamber, with the lead wires in line with the recess in the chamber wall. Push the wiring grommet into the recess. Insert the securing screws and tighten them down evenly. On SR500 models the stator plate has elongated mounting holes, through which the screws pass, to enable alterations to be made in the ignition timing. If on dismantling two index marks were made on the stator plate and adjacent portion of the crankcase, these marks should be aligned exactly before tightening the screws. If the marks were not made ignition timing should be carried out after reassembly of the engine as described in Chapter 3, Section 9.

2 Secure the alternator wires by means of the clips provided on the gearbox wall and reconnect the neutral indicator switch and lead. The terminal should be protected by refitting the rubber boot.

3 Install the Woodruff key in the keyway in the tapered portion of the crankshaft end. Position the alternator rotor over the end of the crankshaft so that the rotor internal keyway aligns with the Woodruff key, and then push the rotor fully home on the shaft. Secure the rotor by means of the nut and washer. To prevent the crankshaft rotating during this operation use the locking technique adopted during dismantling.

35 Reassembling the engine/gearbox unit: replacing the gearchange mechanism

1 Check that the pawl arm is secured correctly on the pivot shaft projecting from the rear of the gearchange quadrant and ensure that the pawl arm return spring is located securely. Slide the quadrant onto the projecting end of the front selector fork rod, depressing the pawl arm at the same time so that it clears the pin and plate at the outer end of the change drum. Fit the quadrant retaining 'E' clip.

2 Fit the gearchange shaft centraliser spring onto the shaft so that one ear of the spring lies each side of the stop bolt (adjuster bolt) on the change arm. The spring must be fitted as shown in the accompanying photograph. Grease lightly the splined end of the shaft so that when it passes through the tunnel in the crankcase it does not damage the lip of the oil seal in the left-hand casing. Insert the gearchange shaft into the tunnel mouth and push it fully home so that the spring ears engage either side of the centraliser anchor bolt in the casing. The main change arm on the shaft must be meshed with the quadrant so that the aligning punch mark on each is opposite the other. Secure the

gearchange shaft by means of the washer and circlip on the left-hand end.

3 Insert the detent plunger, detent spring and housing into the tunnel in the base of the gearbox. Check that the sealing washer on the housing bolt is in good order. Temporarily refit the gearchange lever and attempt to select each gear in turn, starting in the first gear position. Rotate the gearshafts to aid selection. If gear changing is found to be impossible, an error in assembly is indicated. The assembly procedure will have to checked carefully.

4 Place the gearbox in the second gear position and then check that the alignment marks on the change pawl arm and the face of the change drum register (see Fig. 1.13). If the marks are not aligned correctly, slacken the adjuster locknut on the gearchange arm and rotate the adjuster screw, as required, to bring the marks together. Prevent further rotation of the screw and tighten the locknut. The nut is secured by a tab washer which should be bent down before loosening the nut, and bent up again against one of the nut flats after adjustment has been accomplished.

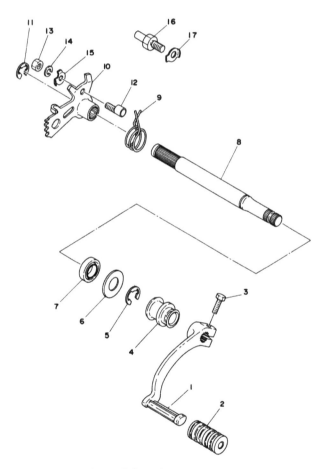

Fig. 1.12 Gearchange mechanism

1	Gearchange lever	10	Gearchange arm
2	Rubber	11	'E' clip
3	Bolt	12	Screw
4	Boot	13	Nut
5	'E' clip	14	Spring washer
6	Washer	15	Lock washer
7	Oil seal	16	Stopper screw
8	Gearchange shaft	17	Lock washer
9	Spring		

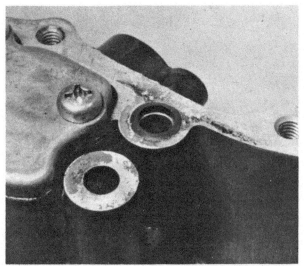

33.1 Install a new O ring in the casing, behind the blanking plate

33.2 Lower the right-hand casing into place squarely

34.1 Position the stator in the casing recess and fit the screws

34.2a Secure the generator leads using the clip(s) provided and ...

34.2b ... reconnect the neutral indicator switch lead

34.3a Install the Woodruff key in the tapered shaft

34.3b Refit the generator rotor and fit and tighten the nut

35.1 Fit the gearchange quadrant, securing it with the E clip

35.2a The gearchange arm centraliser spring must be fitted as shown

35.2b Insert the shaft so that arm and quadrant teeth align correctly

35.2c Secure the gearchange shaft with the washer and E clip

35.3 Insert the detent plunger, spring and housing

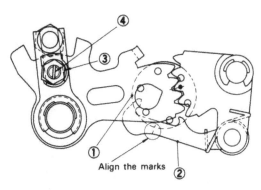

Fig. 1.13 Gearchange selector adjustment

1	*Change drum*	3	*Adjuster locknut*
2	*Pawl arm*	4	*Eccentric adjuster*

36 Reassembling the engine/gearbox unit: replacing the kickstart assembly and oil pump

1 Refit the kickstart stop plate and pawl guide into the primary drive chamber. Secure both plates with the three bolts and then bend up the ears of the locking plate to retain the bolts. If the kickstart shaft assembly was dismantled for examination or renovation, it should be rebuilt as a completed sub-assembly before being inserted into the casing.

2 Slide the completed sub-assembly into place in the casing. Fit the return spring guard plate and then slide the return spring over the shaft so that the inner turned end engages with the radial hole drilled in the shaft. Replace the spring guide on the shaft, pushing it fully home inside the spring. Secure the guide by means of the circlip. Rotate the kickstart shaft as far as possible in a clockwise direction, and then grasp the outer turned end of the spring with a pair of pliers. Tension the spring in a clockwise direction until the outer turned end can be slipped over the anchor lug projecting from the wall of the casing. Finally, slide the heavy thrust washer onto the kickstart shaft.

3 Before replacing the oil pump, check that the pump itself is absolutely clean and that the housing in the crankcase into

which the scavenge pump fits is similarly spotless. Insert the scavenge pump outer rotor into the housing in the casing so that the triangular index mark is facing inwards. Lubricate the rotor thoroughly with clean engine oil and likewise lubricate the main (delivery) pump rotor through the orifices in the pump casing. Insert the oil pump assembly into position so that the scavenge pump inner rotor enters the outer rotor. To align the oil pump casing dowels with the holes in the crankcase rotate the oil pump spindle as required. Fit and tighten the retaining screws only after the pump casing is correctly located by the dowels. Rotate the spindle as the returning screws are tightened, to check for rotor binding. This is usually caused by uneven tightening or dirt on the rotors.

4 Fit the washer, oil pump driven pinion and the second washer onto the oil pump spindle, securing these components with the circlip. The oil pump pinion boss should face towards the pump.

5 Before replacing the kickstart idler pinion onto the projecting end of the gearbox layshaft, slide the heavy thrust washer onto the clutch shaft. Fit the idler pinion and secure it by means of the washer and circlip.

36.1a Fit the kickstart stop plate and ...

36.1b ... the pawl guide and locking plate

36.2a Insert the kickstart shaft sub-assembly into the casing

36.2b Fit the spring guard plate and the return spring

36.2c Insert the guide and tension the spring in a clockwise direction

36.2d Secure the guide with the circlip and ...

36.2c ... install the heavy thrust washer

36.3a Fit the scavenge pump outer rotor and lubricate it thoroughly

36.3b Install the assembled oil pump, aligning the locating dowels

36.4a Fit the oil pump driven pinion and ...

36.4b ... secure it with the washer and circlip

36.5a Replace the heavy thrust washer on the clutch shaft before ...

36.5b ... fitting and securing the kickstart idler gear

37 Reassembling the engine/gearbox unit: replacing the clutch

1   Lubricate the clutch spacer and fit it to the clutch shaft, followed by the kickstart driven gear pinion. This should be placed so that the two drive dogs are projecting outwards. Place the oil pump drive pinion on the boss to the rear of the clutch buter drum so that the two diametrically opposed recesses in the bore align with the recesses in the boss. Lift up the outer drum and pinion and slide them onto the clutch shaft. The dogs projecting from the kickstart driven pinion and the recesses in the rear of the clutch drum must engage with each other. Furthermore the oil pump drive pinion must engage with the oil pump driven pinion. To accomplish this operation some careful manipulation may be required.

2   Fit the heavy thrust washer onto the clutch shaft followed by the clutch centre, belville washer and the centre nut. The washer should be fitted with the concave face inwards. To prevent the clutch rotating while the centre nut is tightened, the rear sprocket should be fitted temporarily and top gear engaged. The sprocket should then be held in position, using the method adopted during dismantling. After tightening the centre nut, grease the clutch pushrod and insert it into the clutch shaft. Insert the 'mushroom' headed thrust piece.

3   Replace the clutch plates one at a time, starting with a friction plate and then a plain plate. Continue inserting the plates

alternatively. Position the clutch pressure plate against the outer friction plate, so that the arrow marks on the pressure plate align with similar marks on the outer face of the clutch centre. Replace the clutch springs and bolts, tightening them down evenly, in a diagonal sequence to maintain a uniform pressure on the plate. The bolts should be tightened fully.

38 Reassembling the engine/gearbox unit: replacing the camshaft drive assembly and primary drive pinion

1   Feed the cam chain up from the primary drive chain through the tunnel mouth to the right of the cylinder, so that it loops around the end of the crankshaft. Slide the cam drive sprocket onto the crankshaft and mesh it with the chain. Prevent the chain from falling back into the casing by securing it with a length of wire. Align the cam chain sprocket recess with the keyway in the crankshaft and insert the long Woodruff key.

2   Slide the timing index plate onto the crankshaft with the timing marks outermost and engage it with the key. Follow suit with the primary drive pinion and contact breaker shaft drive pinion (plain collar on SR500 models). Fit the washer and nut. The nut may be tightened after placing a close fitting bar through the small-end eye, and allowing it to bear down on two wooden blocks placed across the crankcase mouth, to lock the engine.

3   Check that the shim is on the inner portion of the contact breaker cam drive shaft and insert the driveshaft/pinion unit into the casing. This pinion and the pinion on the crankshaft end are each provided with a timing dot on the outer face. In order for ignition timing to be accurately carried out these punched marks must be aligned as the teeth go into mesh. After the alignment has been set the crankshaft **should not be turned** until the camshaft has been refitted and the valve timing completed. If this precaution is not taken, the ignition timing may be 180° out.

4   Position the cam chain tensioner blade so that its bowed face is towards the front of the engine and the locating eye is aligned with the pivot shaft hole in the casing. Insert the pivot shaft so that it locates with the blade eye, and then push it fully home. Do not omit the 'O' ring on the shaft head.

5   Fit a new gasket to the mating surface of the primary drive chamber, locating it on the two hollow dowels. Check that the oil seals in the primary drive cover are in position and lubricate their sealing lips with clean engine oil. Additionally, lubricate the various gears within the engine casing. Position the primary drive cover over the various shafts and push it fully home.

6   Insert and tighten the casing screws evenly, in a diagonal sequence, to help prevent distortion of the case. Do not omit the various cable clamps.

37.1a Place the clutch sleeve and kickstart driven pinion on the shaft

37.1b Align the oil pump pinion slots with those of the clutch boss

37.1c Place the clutch outer drum on the shaft and insert ...

37.1d ... the clutch pushrod

37.2a Replace the clutch centre

37.2b Fit the belville washer with the concave face inwards

37.2c Grease and insert the clutch thrust piece

37.3a Insert the clutch plates one at a time, alternately

37.3b Arrow marks should be aligned

37.3c Tighten the spring bolts evenly

38.1a Pass the cam chain up through the tunnel

38.1b Insert the Woodruff key to locate the cam sprocket

38.2a Fit the timing plate with the marks outermost

38.2b Install the primary drive pinion and ...

38.2c ... the contact breaker drive pinion

38.3 When fitting the contact breaker shaft/pinion the marks must be aligned

38.4 Insert the pivot shaft to locate with the chain tensioner blade

39 Reassembling the engine/gearbox unit: replacing the ATU and contact breaker assembly – except SR500 models

1   Insert the drive pin in the radially drilled hole in the contact breaker driveshaft. Slide the ATU into place on the shaft and rotate it until the pin engages in the recess in the rear boss of the ATU.

2   Position the contact breaker assembly plate in the chamber so that the wiring lead grommet can be pushed into the recess in the chamber wall. Insert the two timing adjustment screws, and screw them in partially. Fit and tighten the ATU centre bolt and washer. If, on dismantling, punch marks were made to aid correct positioning of the contact breaker baseplate, the punch marks should be aligned carefully and the screws tightened fully. The ignition timing should be correct. It is, however, advisable to check the timing as a matter of course, after completion of engine reassembly. If marks were not made, the ignition timing should be carried out after engine reassembly, as described in Chapter 3 Section 9.

39.1 Position the ATU on the shaft so that it locates the drive pin

39.2 Install the complete contact breaker assembly in the casing

40 Reassembling the engine/gearbox unit: replacing the oil filter and oil strainer screen

1   Install a new oil filter element in the filter chamber, so that the raised face of the element locates with the recess in the rear wall of the chamber. Insert a new 'O' ring into the recessed portion of the chamber cover lower mounting bolt hole, and check that the chamber cover 'O' ring is also in place. Refit the chamber cover so that the small bleed screw is in the 12 o'clock position. The shorter of the three chamber screws, which also has a slightly larger diameter shank, should be fitted in the lower screw hole.

2   Tip the engine unit over to one side so that access to the crankcase base is given. Slide the oil strainer screen into the locating groove in the sump cover, and together with a new gasket, fit the cover to the crankcase. Insert and tighten evenly the cover screws. Check that the drain plug is in position and tightened correctly.

41 Reassembling the engine/gearbox unit: replacing the piston, cylinder barrel and cylinder head

1   Place the crankcase unit upright on the workbench, taking care that the camshaft drive chain does not slide back into the crankcase. Fit the piston to the connecting rod, with the arrow on the piston crown facing forwards.

2   The gudgeon pin should be a light sliding fit. If it proves a tight fit, warm the piston first, to expand the metal around the gudgeon pin bosses. Use new circlips to retain the gudgeon pin, and double check to ensure that each is correctly located in the piston boss groove. If a circlip works loose, it will cause serious engine damage.

3   It is advisable to temporarily pad the crankcase mouth during this operation. If a circlip is misplaced, it may fall into the crankcase and necessitate a further engine strip for its retrieval. Leave the rag in position until the cylinder barrel is lowered into position since a similar catastrophe can occur should the piston rings break as they enter the cylinder bore.

4   Coat the cylinder barrel with oil and fit a new base gasket to the crankcase mouth (no cement). Slide a new cylinder base 'O' ring onto the cylinder spigot. Do not omit the four dowels that fit into the base. Slide the cylinder barrel down the holding down studs and gently ease the piston and rings into the bore. There is a generous lead-in on the cylinder base spigot, which should aid this operation. Try to space the piston ring gaps so that they are not in line with each other and preferably at 120° spacings.

5   Whilst the cylinder barrel is lowered into position, the camshaft drive chain should be fed through the tunnel cast in the left-hand side and retained where it emerges at the top, so that it cannot fall back into the crankcase as reassembly proceeds.

6   When the cylinder barrel has seated correctly, fit the three screws through the right-hand base flange, and the four special sleeve nuts. Tighten the four bolts down first, in an even, diagonal sequence, to a torque setting of 3.5-4.0 kgf m (25-29 lbf ft). The screws should be tightened fully, if possible, to a torque setting of 0.8-1.2 kgf m (6-9 lbf ft).

7   Check that the four dowels are in the top of the cylinder barrel and fit a new cylinder head gasket. Lower the cylinder head into position, whilst feeding the cam chain up through the tunnel. When the chain emerges, secure it with a length of stiff wire or by passing a rod through the chain loop.

8   Insert the two socket screws which pass down through the fins, either side of the cam chain tunnel, and fit the washer and domed nut to the stud adjacent to the sparking plug. Do not tighten fully either the screws or nut at this stage.

41.1 The arrow on the piston crown must face forwards when ...

41.2a ... the piston is fitted and the gudgeon pin inserted

41.2b Install **new** circlips, ensuring their correct engagement

41.4a Do not omit to fit a new O ring on the cylinder spigot

41.4b Feed the piston into the cylinder and ...

41.5 ... draw the cam chain up through the tunnel

41.6 Tighten these screws after the four nuts

41.7a Fit a new gasket and the four dowels

41.7b Feed the cam chain through when lowering the head

42 Reassembling the engine/gearbox unit: replacing the overhead camshaft and timing the valves

1   Remove the large hexagon headed inspection plug from the front of the primary drive case so that the index marks on the timing index plate may be seen. Rotate the crankshaft the smallest amount necessary to bring the T marked line in alignment with the index pointer projecting from the casing. In this position the piston is exactly at TDC.

2   Position the camshaft, complete with both bearings into the bearing housings on the cylinder head. Make sure that the bearings are fully seated and that the location circlip on the right-hand bearing is correctly positioned in the housing groove.

3   Insert the camshaft sprocket drive pin into the end of the camshaft and then rotate the camshaft until the pin is in the 12 o'clock position. Position the camshaft sprocket to the right of the camshaft so that the drive pin hole in the rear of the sprocket is in line with the pin and the two scribe lines of the sprocket outer face are parallel with the cylinder head mating surface. Mesh the chains with the sprocket and then push the sprocket onto the camshaft end boss so that the drive pin and drive pin hole engage.

4   Before continuing, check that the crankshaft has not moved – the T mark should still be in line with the pointer – and that

the scribed lines are parallel with the mating surface. If both conditions apply, the valve timing is correct. Fit the sprocket retaining bolt and washer, or on those machines provided with the starting aid for the elderly or infirm, fit the retaining bolt and piston position indicator plate. The plate should engage with the sprocket drive plate, to ensure that it is in the correct relative position. Prevent the camshaft from rotating by applying a spanner to the hexagonal portion of the shaft, and tighten the bolt fully. The correct torque figure is 4.5-5.5 kgf m (33-40 lbf ft).

5   Check that the cylinder head upper mating surface and that of the rocker box are absolutely clean. Apply a good quality jointing compound to both surfaces. Some care should be taken when applying the compound because no gasket is used at this joint and therefore an oil tight joint is more difficult to obtain. Insert the rocker box end plug into the cut away in the cylinder head. On those models fitted with a piston position indicator, the small alignment mark on the plug should be aligned with the cylinder head mating surface. This will ensure that the inspection window is correctly positioned.

6   Check that the locating dowels are in position and then slide the rocker box into place on the studs. Fit the retaining nuts and washers and the remaining socket screws. Tighten the cylinder head nuts and screws evenly a little at a time following the sequence shown in Fig.1.1. Do not omit to tighten the two screws and the domed nut replaced after the cylinder head was refitted. The correct torque settings for the different screws and nuts are as follows:

Nos 1-5 (10 mm nuts) 3.5-4.0 kgf m (25-29 lbf ft)
Nos 6,7 and 10 (8 mm nuts) 1.8-2.2 kgf m (13-16 lbf ft)
Nos 8,9 11 and 12 (6 mm screws) 0.8-1.2 kgf m (6-9 lbf ft)

7   Assemble the components which make-up the chain tensioner unit and fit them into the rear of the cylinder barrel. Rotate the engine forwards a few revolutions and then align the T mark on the flywheel generator rotor with the index mark on the casing. By following this procedure all the chain slack is placed on the rear run of the chain. Turn the adjuster screw hexagonal head in or out until the face of the hexagon is flush with the end of the pushrod. Tighten the locknut. The chain tensioner adjustment should now be correct. Before replacing the adjuster cover, and soon after the engine is run for the first time, the adjustment should be double-checked. With the engine running at tick-over speed, watch the end of the pushrod. There should be slight movement if the adjustment is correct. If no movement is detected, loosen the adjuster, just enough to promote movement. Fit and tighten the tensioner cover.

8   Refit the rocker feed pipe to the union at the rocker box and

crankcase. Use new sealing washers at the banjo unions to ensure oil tightness. Do not overtighten the banjo bolts as these are hollow and also cross-drilled, and will therefore shear very easily.

9   Before replacing the rocker cover, the valve clearance must be adjusted on each valve. The procedure for each is the same, as follows. Check that the piston is at TDC on the compression stroke; provided that the engine has not been moved since valve timing was carried out the piston will be in the correct position. Check the gap between the valve stem of one valve and the rocker arm end, using a feeler gauge.

If adjustment is required, select the feeler gauge corresponding with the smaller clearance for the valve and slide it into position. Loosen the adjuster locknut and screw the adjuster in or out as required until the feeler gauge is a light sliding fit. Prevent further rotation of the adjuster screw and tighten the locknut fully. Re-check to ensure that the action of tightening the locknut has not altered the clearance. Repeat the operation on the remaining valve. Check that the rocker cover sealing rings are in good condition and are correctly seated and fit the covers. Do not overtighten the two screws holding each cover or distortion may occur.

42.2 Position the camshaft, complete with bearings in the head

42.3a Fit the driven sprocket drive pin

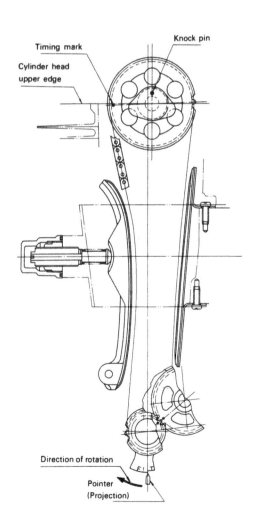
Fig. 1.14 Valve timing

42.3b Align T mark with pointer in the casing

42.3c Drive pin in sprocket should be in 12 o'clock position

42.6a Install the two dowels and end plug before fitting rocker box

42.6b Note decompressor cable anchor retained by screw

42.7a Screw the chain tensioner assembly into the casing

42.7b The large hexagon is the locknut; the small hexagon the adjuster

42.8 Use new sealing washers on the rocker feed pipe unions

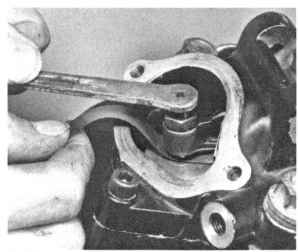

42.9 Feeler gauge should be a light sliding fit

43 Replacing the engine/gearbox unit in the frame

1   Before lifting the engine into the frame, check that the machine is well supported in an upright position, using an arrangement of blocks similar to that adopted during engine removal. As with removal, at least two people should be present when the engine is lifted into place.

2   Lift the engine in from the right-hand side of the machine and lower it into approximately the correct position, supported by the frame tubes. Fit the engine mounting brackets, but do not tighten the nuts and bolts at this stage. Insert the engine bolts from the right-hand side of the engine and fit the nuts. Tighten the engine bracket nuts first and then tighten the main mounting bolts. The head steady bracket may now be fitted. Do not omit the various cables and wiring clips held by the bracket plate upper bolts.

3   Reconnect the exhaust lifter (decompressor) cable with the operating arm on the rocker box. If necessary, adjust the free play in the cable so that there is 5-10 mm (0.2-0.4 in) movement at the ball end of the handlebar lever before the valve begins to lift. This adjustment should take place with the piston at TDC on the compression stroke. Insert the tachometer drive cable in the housing in the rocker box and secure it by means of the internal circlip. If necessary, rotate the engine in order to allow the cable end to align with the driveshaft.

4   Before reconnecting the clutch cable and adjusting the cable free play, the working clearance between the clutch operating shaft and pushrod should be checked and, if necessary, reset. On inspection, it can be seen that the clutch operating arm has a small scribed line marked near the cable anchor, and a small projection protrudes from the crankcase adjacent to the arm. Operate the arm towards the cylinder barrel until all lost motion is taken up and it can be felt that the operating cam has abutted against the pushrod end. In this position the scribed line and projection should be in alignment. If the alignment is incorrect, slacken the large locknut on the adjuster mechanism on the gearbox wall and screw the adjuster screw inwards or outwards until the marks are in line. Tighten

the locknut. The cable may now be passed through the anchor lug in the crankcase and reconnected to the operating arm. To aid reconnection apply an open ended spanner to the operating arm and move it into the operated position. Adjust the clutch cable at the handlebar lever adjuster as described in Routine maintenance.

5   Check that the 'O' ring on the oil return pipe lower union is in position and insert the union into the aperture in the top of the crankcase. Fit the two retaining screws and tighten them down evenly. Note that on some models the front screw also retains a wiring clip. Using a pressure oil gun prime the oil feed passage through the union orifice in the gearbox wall. Insert a new 'O' ring into the recess in the machined face. Slide the oil delivery pipe under the engine so that the gland union is at the front. Secure the pipe by means of the single clamp and screw and secure the flanged union by means of the two socket screws. Do not overtighten the flange screws, otherwise distortion may occur. Before reconnecting the oil pipe with the union at the base of the frame front downtube, prime the pipe using an oil gun. This will aid circulation of oil when the engine is first started.

6   Refit the carburettor to the cylinder head and interconnect it with the air filter hose. If the carburettor mounting stub was removed, a new gasket must be fitted between the flange face and the inlet port during reassembly. Reconnect the throttle cable with the pulley on the carburettor. The opening cable should be fitted to the upper cable anchor. Adjust the cables so that there is about 5 mm (0.2 in) movement at the throttle twist grip before the throttle starts opening.

7   Refit the final drive sprocket to the output shaft. If an endless chain is used, the sprocket should be meshed with the chain before fitting. Fit the sprocket tab washer and the securing nut. Where a detachable chain is used, loop the chain round the engine sprocket and reconnect the two ends with the master link. Ensure that the spring link is correctly fitted with the closed end facing the normal direction of chain travel. With the chain in place the sprocket nut may be tightened. Apply the rear brake and select top gear to prevent rotation of the shaft. Bend up one edge of the tab washer against one of the nut flats to secure the nut. Replace the flywheel generator cover and the final drive sprocket cover.

8   Before replacing the breather chamber at the rear of the gearbox roof, pour approximately 1.0 litre (2.2/1.8 US/Imp pints) into the engine via the breather union projecting from the casing. This will give a sufficient priming quantity of oil to protect the engine during initial starting. The breather chamber should be fitted so that the alignment mark on the chamber lower joint pipe aligns with the similar mark on the left-hand side of the union.

9   Reconnect the various electrical leads and secure them by means of the clips provided. Install the battery (where fitted) and reconnect it, ensuring that the red lead is connected to the positive terminal and the black lead (earth) is connected to the negative terminal.

10 Place a new ring gasket in the exhaust port and reassemble the exhaust system. Replace the rubber mounting sub-components in the sequence shown in the relevant illustrations in Chapter 2. The right-hand suspension strut can now be refitted. Ensure that the upper and lower mounting bolts are tightened fully.

11 Check that all the engine and oil tank drain plugs have been fitted and tightened fully and that the oil lines have been connected. Replenish the oil tank with about 1.5 litre (3.0/2.5 US/Imp pints) of SAE 20W/50 engine oil. The final check on all levels should be made after the first engine start-up, when the oil has circulated fully and dispersed somewhat within the engine.

12 Refit the petrol tank and the dualseat. Reconnect the petrol pipe, and on SR500 models the top vacuum pipe. Do not omit the pipe securing spring clips or the anti-kink springs.

13 Check around the engine to make certain that all components have been fitted and are functioning correctly and that all bolts, nuts and screws are correctly tightened.

43.2a Lift the engine in from the right-hand side

43.2b Detachable engine brackets are fitted both front ...

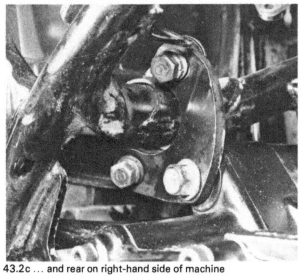

43.2c ... and rear on right-hand side of machine

43.2d The head steady bracket bolt should be tightened last

43.4 Secure the clutch cable nipple by means of the small tang

43.5a Check condition of O ring in return pipe and ...

43.5b ... the oil feed pipe unions

43.7a Tighten the final drive sprocket nut and then ...

43.7b ... bend up lock washer to secure the nut

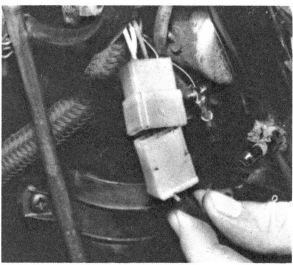

43.9a Reconnect the electrical lead block connectors and ...

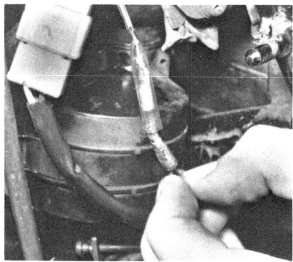

43.9b ... any single snap connectors

43.11 Refill the oil tank with the correct quantity and specification of oil

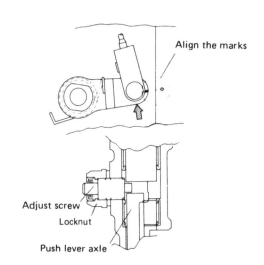

Fig. 1.15 Clutch arm adjustment marks

lubrication failure investigated.

3   Check the movement of the cam chain tensioner plunger rod as described in paragraph 7 of the previous section. Make any required adjustments and then fit the tensioner cap.

4   The engine may show a tendency to smoke initially due to the amount of oil used during assembly of the various components. The excess of oil should gradually burn away as the engine settles down.

5   Check the exterior of the machine for oil leaks or blowing gaskets. Make sure each gear engages correctly and that all controls function effectively, particularly the brakes. This is an essential last check before taking the machine on the road.

45 Taking the rebuilt machine on the road

1   Any rebuilt machine will need time to settle down, even if parts have been replaced in their original order. For this reason it is highly advisable to treat the machine gently for the first few miles to ensure oil has circulated throughout the lubrication system and that any new parts fitted have begun to bed down.

2   Even greater care is necessary if the engine has been rebored or if a new crankshaft has been fitted. In the case of a rebore, the engine will have to be run in again, as if the machine were new. This means greater use of the gearbox and a restraining hand on the throttle until at least 500 miles have been covered. There is no point in keeping to any set speed limit; the main requirement is to keep a light loading on the engine and to gradually work up performance until the 500 mile mark is reached. These recommendations can be lessened to an extent when only a new crankshaft is fitted. Experience is the best guide since it is easy to tell when an engine is running freely.

3   If at any time a lubrication failure is suspected, stop the engine immediately and investigate the cause. If an engine is run without oil, even for a short period, irreparable engine damage is inevitable.

4   When the engine has cooled down completely after the initial run, recheck the various settings, especially the valve clearances. During the run most of the engine components will have settled down into their normal working locations.

44 Starting and running the rebuilt engine

1   Open the petrol tap to allow fuel to flow to the carburettor. On SR500 models the tap lever should be placed in the 'Prime' position, to allow an unrestricted flow of fuel. When the engine has started the lever may be returned to the 'On' position and the fuel flow will be controlled by the inlet vacuum in the normal way.

2   Remove the oil bleed screw from the top of the oil filter chamber cover. After a few moments of engine running, the chamber should fill with oil, which will then be ejected forcibly through the bleed orifice. As soon as this happens stop the engine and refit the bleed screw. If no oil appears after a few seconds running, the engine **must** be stopped and the cause of

46 Fault diagnosis: engine

Symptom	Cause	Remedy
Engine will not start	Defective spark plug	Remove the plug and lay it on cylinder head. Check whether sparking occurs when ignition is switched on and engine rotated.
	Dirty or closed contact breaker points (Except SR500 model)	Check condition of points and whether gap is correct.
	Faulty or disconnected condenser (Except SR500 model)	Check whether points arc when separated. Replace condenser if evidence of arcing.
	Faulty CDI unit (SR500 models)	Check unit for electrical resistance. Renew if faulty.
	Faulty CDI source coil or pulser (SR500 models)	Check circuits. Renew components if faulty.
Engine runs unevenly	Ignition and/or fuel system fault	Check each system independently, as though engine will not start.
	Blowing cylinder head gasket	Leak should be evident from oil leakage where gas escapes.
	Incorrect ignition timing	Check accuracy and if necessary reset.
Lack of power	Fault in fuel system or incorrect ignition timing	See above.

Heavy oil consumption	Cylinder barrel in need of rebore	Check for bore wear, rebore and fit oversize piston if required.
	Damaged oil seals	Check engine for oil leaks.
Excessive mechanical noise	Worn cylinder barrel (piston slap)	Rebore and fit oversize piston.
	Worn camshaft drive chain (rattle)	Adjust tensioner or replace chain.
	Worn big end bearing (knock)	Fit replacement crankshaft assembly.
	Worn main bearings (rumble)	Fit new journal bearings and seals.
Engine overheats and fades	Lubrication failure	Stop engine and check whether internal parts are receiving oil. Check oil level in oil tank.

47 Fault diagnosis: clutch

Symptom	Cause	Remedy
Engine speed increases as shown by tachometer but machine does not respond	Clutch slip	Check clutch adjustment for free play at handlebar lever. Check thickness of inserted plates.
Difficulty in engaging gears. Gear changes jerky and machine creeps forward when clutch is withdrawn. Difficulty in selecting neutral	Clutch drag	Check clutch adjustment for too much free play. Check clutch drums for indentations in slots and clutch plates for burrs on tongues. Dress with file if damage not too great.
Clutch operation stiff	Damaged, trapped or frayed control cable	Check cable and replace if necessary. Make sure cable is lubricated and has no sharp bends.

48 Fault diagnosis: gearbox

Symptom	Cause	Remedy
Difficulty in engaging gears	Selector forks bent	Replace.
	Gear clusters not assembled correctly	Check gear cluster arrangement and position of thrust washers.
Machine jumps out of gear	Worn dogs on ends of gear pinions	Replace worn pinions.
	Stopper arms not seating correctly	Remove right hand crankcase cover and check stopper arm action.
Gearchange lever does not return to original position	Broken return spring	Replace spring.
Kickstarter slips	Ratchet assembly worn	Part crankcase and replace all worn parts.
Kickstarter does not return when engine is turned over or started	Broken or poorly tensioned return spring	Replace spring or re-tension.

Chapter 2 Fuel system and lubrication

Refer to Chapter 7 for information relating to the 1979 to 1983 models

Contents

Specifications

Fuel tank

	TT500C, D and E,XT500E	XT500C and D	SR500
Capacity	8.5 lit (2.25/1.87 US/Imp galls)	8.8 lit (2.32/1.94 US/Imp galls)	12 lit (3.1/2.64 US/Imp galls)

Carburettors

	TT500E and XT500E (US)	TT500C and XT500C	TT500D	XT500D and E (UK)	SR500
Make	Mikuni	Mikuni	Mikuni	Mikuni	Mikuni
Type	VM32SS	VM34SS	VM34SS	VM32SS	VM34SS
Main jet	230	210	240	220	300
Air jet	45.	0.8	0.8	0.8	80
Jet needle	6FL24-3	5J10-3	6H2-4	6H2-4	6FL25-2
Needle jet	Q-0	Q-2	Q-2	Q-2	P-8
Cut-away	3.5	4.5	4.0	4.0	3.5
Pilot jet	25	35	30	25	25
Starter jet	55	60	60	60	50
Air screw (turns out)	preset	$1\frac{1}{4}$ ($1\frac{3}{4}$XT500C)	$1\frac{3}{4}$	$1\frac{1}{4}$	preset

Oil capacity

Dry	2.5 lit (5/4.4/US/Imp pints)
Without filter change	2.0 lit (4.2/3.5 US/Imp pints)
With filter change	2.1 lit (4.4/3.7 US/Imp pints)

Oil pump

Type	Trochoid
Housing maximum depth:	
Delivery (main)	4.09 mm (0.1610 in)
Scavenge	18.09 mm (0.7122 in)
Housing maximum ID	40.85 mm (1.6083 in)
Rotor minimum OD	40.50 mm (1.5945 in)
Rotor minimum thickness:	
Delivery	3.95 mm (0.1555 in)
Scavenge	17.95 mm (0.70669 in)
Outer rotor/housing clearance	0.09-0.15 mm (0.0035-0.0059 in)
Wear limit	0.35 mm (0.0138 in)
Outer rotor/inner clearance	0.07-0.12 mm (0.0028-0.0047 in)
Wear limit	0.35 mm (0.0138 in)

1 General description

The fuel system comprises a petrol tank astride the top frame tube from which petrol is fed to the Mikuni carburettor. A lever-type petrol tap with a detachable gauze filter and filter bowl, is located at the rear left-hand underside of the petrol tank. It has a reserve position, to permit the machine to be ridden a short distance after the main fuel supply has run out. The SR500 models are fitted with a diaphragm type petrol tap which is controlled by the vacuum in the inlet tract.When the tap is in the 'On' or 'Res' position fuel will flow only when the engine is running. If no fuel is in the carburettor, for instance after the carburettor has been dismantled and reassembled, the petrol tap may be placed in the 'Prime' position so allowing an unrestricted flow of petrol. The tap lever should be returned to the 'On' position as soon as the engine is running.

A large capacity air cleaner, with a detachable oil-soaked plastic foam element is mounted on the right-hand side of the machine, within a moulded plastic box. It is attached direct to the carburettor intake by a short rubber hose. To aid starting from cold, the carburettor has a choke for temporarily richening the mixture. On SR500 models an aid to warm starting is provided also. This consists of a small button, integral with the throttle stop screw, which may be depressed prior to starting and so places the throttle slide that the correct mixture for starting is automatically provided without the need for manual operation of the throttle twist grip. As soon as starting has been successfully accomplished and the throttle is moved, the button returns to its normal un-set position. Lubrication is provided by a dry sump system where the oil is contained in the frame top and front downtubes. A double trochoid oil pump is used to deliver oil to the working surfaces of the engine/gearbox and to return the circulated oil to the oil tank. The pump consists of two separate rotor sets mounted axially upon the same driveshaft. The engine is protected by a paper element oil filter and two gauze oil strainer screens. One screen is fitted to the oil feed union in the oil tank and the other is fitted within the crankcase sump.

2 Petrol tank: removal and replacement

1   Before the petrol tank is removed, the dualseat must be detached by unscrewing the two bolts which pass through seat base lugs and into the frame.
2   On SR500 models the petrol tank is retained on the frame by a single bolt that passes through a lug welded to the extreme rear end of the tank. The nose of the tank locates with a rubber buffer on each side of the steering head by means of guide channels welded to the inside of each overhanging section. To remove the tank, disconnect the petrol pipe and vacuum hose from the petrol tap. Unscrew the bolt at the rear of the tank and ease the tank backwards until the locating cups at the front leave the rubbers. It can then be lifted clear of the frame.
3   All remaining models are fitted with a petrol tank which is secured by a single bolt at the rear and one bolt either side at the front. Each bolt passes through a lug, isolated by a rubber damper. After disconnecting the petrol pipe and removing the bolts, the tank may be lifted up and away from the machine.
4   Refit the petrol tank by reversing the removal procedure. Ensure that any cables or wires which run below the tank are not trapped or broken during replacement.

3 Petrol tap: removal, dismantling and replacement

1   At regular intervals the petrol tap should be removed from the tank to enable the fuel filter to be cleaned and inspected. Before the petrol tap can be removed, it is first necessary to drain the tank. This is easily accomplished by removing the feed pipe from the carburettor float chamber and allowing the contents of the tank to drain into a clean receptacle. Place the tap lever in the 'Res' position, or where a diaphragm controlled tap is used in the 'Prime' position to allow the fuel to flow.
2   Depending on the model, the tap is retained on the underside of the tank either by a gland nut concentric with the tap body, or by two screws passing through the tap flange. In the latter case an 'O' ring is interposed between the flange and the tank surface to effect a petrol-tight seal. Remove the tap and clean the filter in petrol, if necessary agitating any adhering foreign matter with a soft brush. The filter screen should be renewed if it has been perforated. When replacing the tap, check that the sealing ring (where fitted) is in good condition and is seated correctly in the flange groove. To aid sealing of the tap where a gland nut fixing is utilised, apply a small amount of petrol resistant gasket compound to the threads before refitting. Some taps are fitted with a filter bowl which screws into the base of the tap and incorporates a small filter. This acts as a sediment trap, and should be unscrewed and flushed out during filter screen cleaning. On SR500 models, a sediment trap is fitted which is held to the underside of the tap by four screws. This should be removed at intervals, for cleaning.
3   If the tap fitted to SR500 and XT 500 models leaks there is no necessity to remove the main body of the tap from the petrol tank, even though there is still need to drain the tank. If the two screws in the lever surround are withdrawn, the complete lever assembly can be taken out and the packing behind the lever removed. The tap fitted to TT500 models is a sealed unit. If the tap leaks at the lever, the complete unit should be renewed.

2.2a The petrol tank is retained by a single bolt at the rear and ...

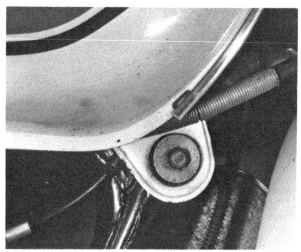
2.2b ... two bolts passing through lugs at the front

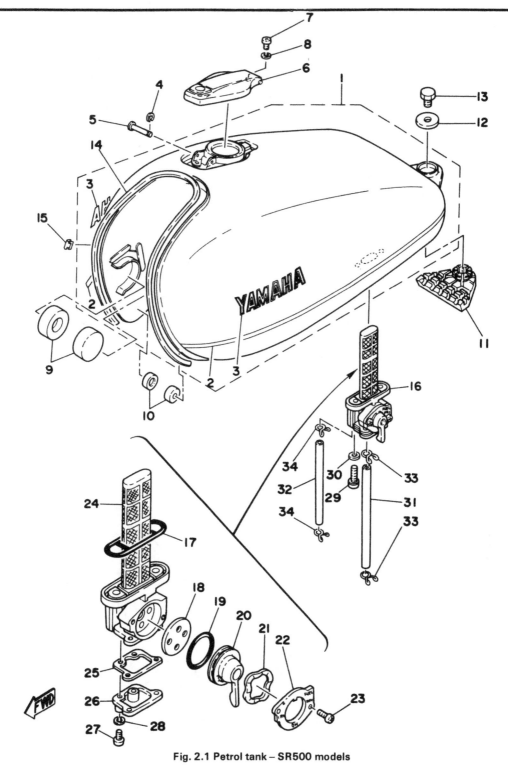

Fig. 2.1 Petrol tank – SR500 models

1 Petrol tank	13 Bolt	24 Filter column
2 Lining strip	14 Rubber moulding	25 Gasket
3 Emblem	15 Trim clip – 6 off	26 Lower cover
4 'E' clip	16 Petrol cap assembly	27 Screw – 4 off
5 Hinge pin	17 Sealing ring	28 Spring washer – 4 off
6 Tank cap	18 Valve plate	29 Screw – 2 off
7 Screw	19 'O' ring	30 Sealing washer – 2 off
8 Spring washer	20 Tap lever	31 Petrol pipe
9 Mounting rubber – 2 off	21 Wave washer	32 Vacuum pipe
10 Damper rubber – 2 off	22 Outer cover	33 Spring clip – 2 off
11 Rubber seat	23 Screw – 2 off	34 Spring clip – 2 off
12 Plain washer		

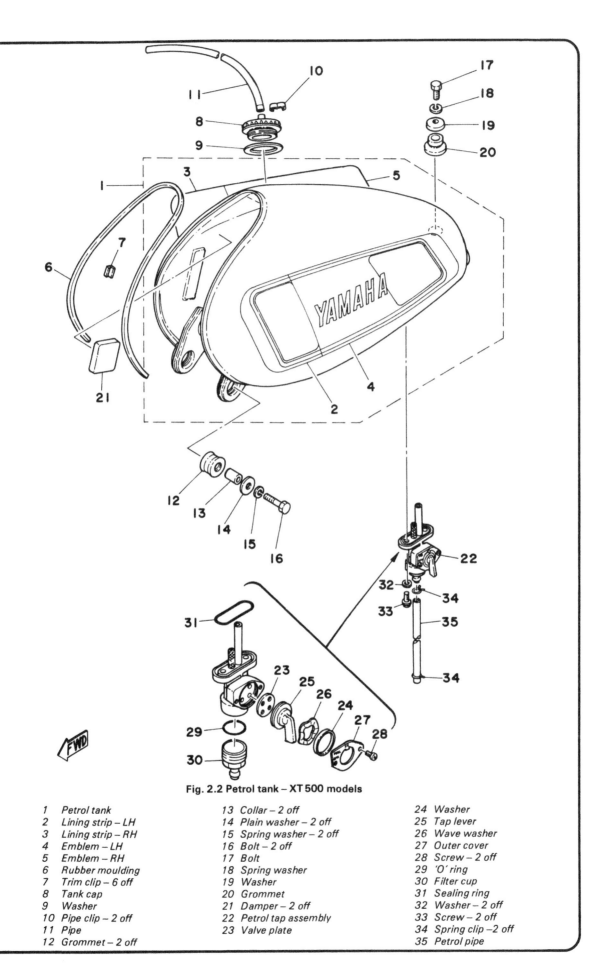

Fig. 2.2 Petrol tank – XT 500 models

1	Petrol tank	13	Collar – 2 off	24	Washer
2	Lining strip – LH	14	Plain washer – 2 off	25	Tap lever
3	Lining strip – RH	15	Spring washer – 2 off	26	Wave washer
4	Emblem – LH	16	Bolt – 2 off	27	Outer cover
5	Emblem – RH	17	Bolt	28	Screw – 2 off
6	Rubber moulding	18	Spring washer	29	'O' ring
7	Trim clip – 6 off	19	Washer	30	Filter cup
8	Tank cap	20	Grommet	31	Sealing ring
9	Washer	21	Damper – 2 off	32	Washer – 2 off
10	Pipe clip – 2 off	22	Petrol tap assembly	33	Screw – 2 off
11	Pipe	23	Valve plate	34	Spring clip –2 off
12	Grommet – 2 off			35	Petrol pipe

3.2a Remove the tap to allow cleaning of the filter

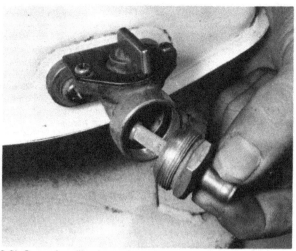

3.2b Secondary filter is used on some types of tap

3.3a Remove the two small screws and ...

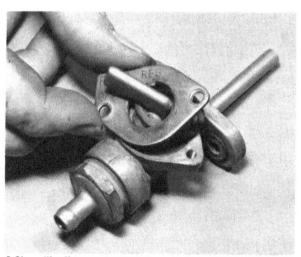

3.3b ... lift off cover plate to allow ...

3.3c ... removal of lever and access to the seal

---

4 Carburettor: removal

1 To improve access to the carburettor, the petrol tank and dualseat should be removed as described in Section 2 of this Chapter.

2 Slacken the adjuster screw at the lower end of each throttle cable after loosening the locknut. Slip one adjuster out of the cable anchor and detach the nipple and inner cable from the control pulley. Repeat the operation for the second cable.

3 Slacken fully the carburettor/air filter hose screw clip and remove the two socket screws which retain the inlet stub flange at the cylinder head. Pull the carburettor out at the front so that the inlet stub clears the cylinder head. The carburettor can then be pulled out of the air hose. If difficulty is encountered in carburettor removal due to the limited space between the air filter box and cylinder head, the air filter box may be removed.

5   Carburettor: dismantling, examination and reassembly

All UK XT500 models, and US XT/TT500 C and D models

1   Remove the two screws which pass through the carburettor cap, and lift the cap away. Bend down the ear of the tab washer which secures the bolt passing through the bellcrank boss. The bolt locates the operating shaft upon which is mounted the throttle control pulley. Remove the bolt and pull out the operating shaft, to free the bellcrank. The bellcrank can be lifted out together with the throttle valve (slide) and jet needle.

2   In normal circumstances further dismantling of the throttle valve/ needle assembly is not required. However, if renewal of the needle is required, continue as follows: remove the two screws which pass through the throttle valve top. This will release the bell crank and link assembly. These two screws are usually very tight and because of their size easily damaged. It follows that care should be taken when loosening them. Push out the jet needle, together with the needle clip.

3   Invert the carburettor and detach the float chamber by removing the four screws and spring washers that retain it to the main body. There is a sealing gasket around the edge of the float chamber, which will remain either with the float chamber or the main body of the carburettor. It need not be disturbed unless it is broken or deformed.

4   Pull out the pivot pin from the twin float assembly and lift the floats away. The float needle will now be displaced from its seating and should be put aside in a safe place for examination at a later stage. It is very small and easily lost. If, on subsequent examination, it is found that the float needle assembly is worn, the valve seat may be removed. It is screwed into place.

5   Apply a small spanner to the main jet holder and then unscrew the main jet. The jet holder can then be unscrewed from position. Note the small locating 'O' ring around the top of the holder. Invert the carburettor to allow the needle jet to fall out.

6   The pilot jet screws into the base of the mixing chamber body to one side of the main jet housing column. This jet may be removed for cleaning.

All SR models, and US XT/TT500 E models onwards

7   The carburettor fitted to these models is similar in most respects to the carburettor described above, but has some additional components. These include a throttle pump, override valve and a valve to richen the mixture when the engine is on the over-run. The procedure for dismantling this instrument is materially the same as that described in paragraphs 1–5 with the following variations and additions:

8   Before removing the throttle operating shaft disconnect the throttle pump link rod from the pulley wheel.

9   If the needle requires removal from the throttle slide take note of the sequence of the spring, clip and upper and lower seats which fit on the needle itself. They should be refitted in the same sequence.

10 The float needle valve seat is a push fit in the float chamber roof, located by a small 'O' ring and secured by a claw plate held by a single screw.

11 Removal of the mixture richening valve from the side of the carburettor body or the throttle pump and override valve from the base of the float chamber is not strictly necessary unless some indication has been given that one of these items has malfunctioned. The valve covers are retained by three screws each and the throttle pump assembly cover by four screws. When releasing the cover from any of these units, note that a spring is fitted to return the valve or pump diaphragm. Steps should be taken to prevent the cover and spring from flying out.

12 The diaphragm of each unit should be checked for perishing or perforation. If these conditions are evident, the diaphragm block in question should be renewed.

All models

13 Check that the floats are in good order and not punctured. Because they are moulded from a plastic material, it is not possible to effect a permanent repair. In consequence, a new replacement should always be fitted if damage is found.

14 The float needle seating will wear after lengthy service and should be closely examined with a magnifying glass. Wear usually takes the form of a ridge or groove, which will cause the float needle to seat imperfectly. Always renew the seating and float needle as a pair, especially since similar wear will almost certainly occur on the point of the needle.

15 After a considerable mileage has been covered, the interaction between the needle jet and jet needle will cause both of these components to wear, resulting in an increased petrol consumption. The two items should be renewed as a pair.

16 The choke, provided for cold starting, is of the plunger type, located in a tunnel in the side of the carburettor body. It is unlikely that malfunction of the choke will occur, due to limited use even after an extended mileage, and in normal circumstances it should not require attention. The plunger unit may be removed after removing the operating lever and pivot screw and unscrewing the plunger retaining cap.

17 Before the carburettor is reassembled, using the reversed dismantling procedure, it should be cleaned out thoroughly using compressed air. Avoid using a piece of rag since there is always risk of particles of lint obstructing the internal passageways or the jet orifices.

18 Never use a piece of wire or any pointed metal object to clear a blocked jet. It is only too easy to enlarge the jet under these circumstances and increase the rate of petrol consumption. If compressed air is not available, a blast of air from a tyre pump will usually suffice.

19 Do not use excessive force when reassembling a carburettor because it is easy to shear a jet or some of the smaller screws. Furthermore, the carburettor is cast in a zinc-based alloy which itself does not have a high tensile strength. If any of the castings are damaged during reassembly, they will almost certainly have to be renewed.

6   Carburettor: adjusting the mixture and tick-over speed

1   Adjustment of the mixture for a regular tick-over speed and adjustment of the tick-over speed itself should be carried out only after the engine has been allowed to reach normal temperature. If the adjustments are made when the engine is cold, the results will be incorrect during normal operation.

2   Screw in the pilot screw until it seats lightly and then screw it out exactly the amount specified for each model. It should be noted, however, that on some models the pilot screw is fitted with a limiter cap. On adjustment at the factory the cap is fitted to prevent injudicious over-adjustment of the screw. Where the screw limiter is fitted do not move the screw at this stage.

TT500C, XT500D and XT500E (UK)	$1\frac{1}{4}$ turns out
TT500D and XT500C	$1\frac{3}{4}$ turns out

Start the engine and allow it to tick-over. Turn the pilot adjuster screw in or out a small amount until the highest rpm is found. On these carburettors with a limiter cap the screw should be moved only within the confines of the limit stops.

3   Using the throttle stop screw with the knurled head adjust the tick-over speed to the following rpm for each model.

SR500, TT500C and XT500D models 1100 rpm
TT500D and E and XT500E models 1200 rpm

Carry out the pilot screw adjustment once more and then, if required, repeat the tick-over adjustment operation.

On those machines not fitted with a tachometer and where an ancillary tachometer is not available for testing purposes the tick-over should be set by ear.

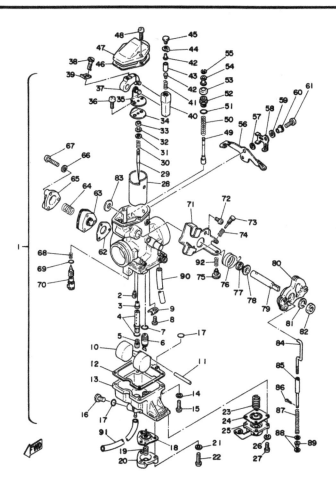

Fig. 2.3 Carburettor – all SR500 models, and US XT/TT500 E models onwards

1 Carburettor assembly	32 Needle clip	63 Cut-off valve
2 Pilot jet	33 Needle collar	64 Spring
3 Needle jet	34 Spring plate	65 Cover
4 Main jet holder	35 Spring seat	66 Spring washer – 3 off
5 Main jet	36 Screw – 2 off	67 Screw – 3 off
6 Float needle assembly	37 Bell crank	68 Spring
7 'O' ring	38 Bolt	69 'O' ring
8 Screw	39 Tab washer	70 Pilot adjuster screw
9 Claw plate	40 Link block	71 Cable anchor plate
10 Float assembly	41 Spring	72 Screw – 2 off
11 Pivot pin	42 Locator pin – 2 off	73 Throttle opening stop screw
12 Gasket	43 Locator rod	74 Spring
13 Float bowl	44 Tab washer	75 Throttle stop screw
14 Spring washer– 4 off	45 Bolt	76 Throttle return spring
15 Screw – 4 off	46 Gasket	77 Seal
16 Drain plug	47 Carburettor top	78 Shouldered washer
17 'O' ring	48 Screw – 2 off	79 Operating rod
18 Release diaphragm	49 Choke plunger	80 Cable pulley
19 Spring	50 Spring	81 Spring washer
20 Cover	51 'O' ring	82 Nut
21 Spring washer – 3 off	52 Choke housing	83 Nylon washer
22 Screw – 3 off	53 Dust cap	84 Throttle pump link
23 Spring	54 Collar	85 Adjuster rod
24 Accelerator pump diaphragm	55 Circlip	86 Split pin
25 Pump cover	56 Choke lever	87 Spring
26 Spring washer – 4 off	57 Plain washer	88 Plain washer – 2 off
27 Screw – 4 off	58 Throttle link	89 Shouldered collar
28 Throttle valve	59 Plain washer	90 Vent pipe
29 Jet needle	60 Shouldered collar	91 Overflow pipe
30 Spring	61 Screw	92 Spring
31 Needle washer	62 Gasket	

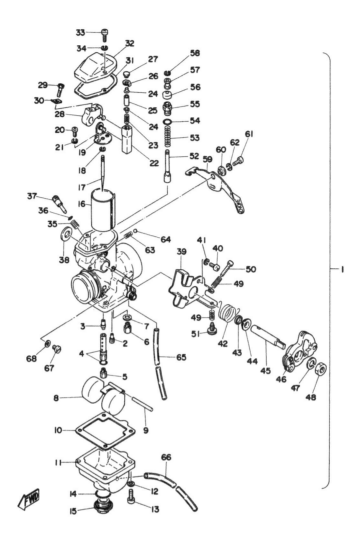

Fig. 2.4 Carburettor assembly – all UK XT models, and US XT/TT500 C and D models

1	Carburettor assembly	22	Link block	43	Seal
2	Pilot jet	23	Spring	44	Shouldered washer
3	Needle jet	24	Locating pin – 2 off	45	Operating shaft
4	Main jet holder	25	Locating rod	46	Cable pulley
5	Main jet	26	Tab washer	47	Spring washer
6	Float needle valve	27	Bolt	48	Nut
7	Sealing washer	28	Bell crank	49	Spring
8	Float assembly	29	Bolt	50	Throttle opening stop screw
9	Pivot pin	30	Tab washer	51	Throttle stop screw
10	Gasket	31	Gasket	52	Choke plunger
11	Float bowl	32	Carburettor top	53	Spring
12	Spring washer – 4 off	33	Screw – 2 off	54	'O' ring
13	Screw – 4 off	34	Spring washer – 2 off	55	Plunger housing
14	'O' ring	35	Spring	56	Dust cap
15	Drain plug	36	'O' ring	57	Collar
16	Throttle valve	37	Pilot adjustment screw	58	Circlip
17	Jet needle	38	Nylon washer	59	Choke lever
18	Clip	39	Cable anchor bracket	60	Shouldered collar
19	Needle plate	40	Screw – 2 off	61	Screw
20	Screw – 2 off	41	Spring washer – 2 off	62	Spring washer
21	Spring washer – 2 off	42	Throttle return spring		

5.1a Bend down the tab washer and remove the bolt

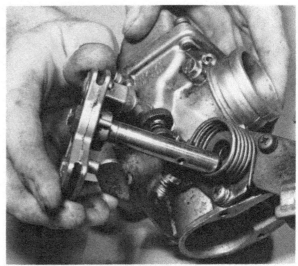

5.1b Withdraw the operating shaft noting ...

5.1c ... the nylon washer between crank and carburettor body

5.1d The bell crank and throttle valve may be withdrawn

5.2 Two screws hold link to throttle valve

5.3a Remove the carburettor float bowl

5.3b Displace the pivot pin to free the float assembly

5.3c Do not lose the tiny float needle

5.4a Unscrew the main jet and then ...

5.4b ... the main jet holder. Note the O ring

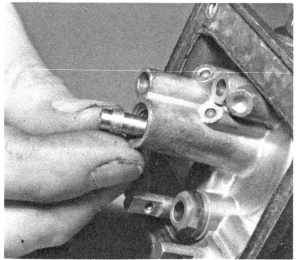

5.4c The needle jet is a push fit in the body and ...

5.4d ... protrudes into the venturi bore

6.2 Location of pilot adjuster screw

7 Carburettor: checking the float chamber fuel level

1 If conditions of a continual weak mixture or flooding are encountered or if difficulty is experienced in tuning the carburettor, the float level should be checked and, if necessary, adjusted. Although the float chamber may be removed with the carburettor in situ on the machine, it is advised that the carburettor be removed to facilitate inspection and adjustment.

2 The float level is correct when the distance between the uppermost edge of the floats (with the carburettor inverted) and the mixing chamber body flange is as follows:

TT500C and D,	*22·0 ± 2·5 mm*
D and E (UK)	*(0·866 ± 0·098 in)*
TT500E and XT500E (US)	*34·0 ± 0·5 mm*
	(1·339 ± 0·020 in)
SR500	*23·5 ± 1·0 mm*
	(0·9173 ± 0·040 in)

The gasket must be removed from the mixing chamber body before the measurement is taken. The floats should be in the closed position when the measurement is taken, with the float tang just touching but not depressing the spring loaded portion of the float needle. Adjustment is made by bending the float assembly tang (tongue), which engages with the float in the direction required.

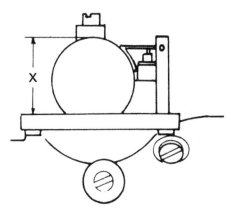

Fig. 2.5 Checking the float level
x = Distance to be measured

8 Carburettor: settings

1 Some of the carburettor settings such as the sizes of the needle jets, main jets and needle positions are pre-determined by the manufacturer. Under normal riding conditions it is unlikely that these settings will require modification. If a change appears necessary it is often because of an engine fault, or an alteration in the exhaust system eg; a leaky exhaust pipe connection or silencer.

2 As an approximate guide to the carburettor settings, the pilot jet controls the engine speed up to $\frac{1}{8}$th throttle. The throttle slide cut-away controls the engine speed from $\frac{1}{8}$th to $\frac{1}{4}$ throttle and the position of the needle in the slide from $\frac{1}{4}$ to $\frac{3}{4}$ throttle. The size of the main jet is responsible for engine speed at the final phase of $\frac{3}{4}$ to full throttle. These are only guide lines; there is no clearly defined demarkation line due to a certain amount of overlap that occurs.

3 Always err slightly towards a rich mixture as one that is too weak will cause the engine to overheat and burn the exhaust valves. Reference to Chapter 3 will show how the condition of the sparking plug can be interpreted with some experience as a reliable guide to carburettor mixture strength.

4 Alterations to the mid-range mixture strength can be made by changing the position of the throttle needle in the throttle slide by moving the needle clip into a different groove. Raising the needle will richen the mixture and lowering the needle will weaken it.

9 Air cleaner: dismantling, servicing and reassembly

1 The air filter box is fitted immediately to the rear of the carburettor, to which it is connected by a short rubber hose. The filter box contains a detachable air filter element which may be removed for cleaning.

2 After removal of the frame right-hand side cover, to gain access to the filter box, the box side cover may be removed. The cover is held by three screws (XT and TT models) or four screws (SR500 models) of which each is fitted with a plain washer and a spring washer. To remove the air filter element on SR500 models depress the spring steel retaining arm and slide the element from position. All other models have a plate at the front and the rear of the element which support the element. Grasp both plates and slide them out simultaneously, together with the filter element.

3 The type of element used, and hence the cleaning method adopted, differs depending on the model. All SR500 models are fitted with a dry fabric mesh element permanently affixed to the element frame. Tap the element firmly to shake off all loose dust, and then blow the element out from the inside. Ideally an air hose should be used, but in the absence of this a tyre pump can be used as a substitute. Do not try and brush the dust off as this will force dust into the fabric, causing blockage. If the element is perforated or badly soiled with lubricant, it should be renewed.

4 All remaining models are fitted with an oil-impregnated foam element. Remove the element from the supporting frame and wash it thoroughly in petrol. Squeeze out the element to remove as much petrol as possible and then leave the element for a short while to allow the remaining petrol to evaporate. Do not wring out the sponge as this will cause damage necessitating early replacement of the element. Re-impregnate the element with SAE 30 engine oil and then squeeze it gently, to remove the excess oil. The element should be wet but not dripping. If the sponge becomes damaged or hardens with age, it should be renewed.

5 On all models refit the element by reversing the dismantling procedure. Where a sponge element is used, the faces of the element frame should be coated with grease to ensure an airtight seal between them and the support plates. Check also that the sealing ring on the front plate is in good condition and is seated correctly.

6   Never run the machine without the element or with the air cleaner disconnected, otherwise the weak mixture that results will cause engine overheating and severe damage.

10 Exhaust system

1   Unlike a two-stroke, the exhaust system does not require frequent attention because the exhaust gases are of a less oily nature and in consequence the internal baffles are much less likely to block. There is no means of detaching the baffles from the silencer to facilitate their cleaning; usually the silencer will

have reached the point of visible deterioration long before there is any chance of obstruction through carbon build-up.

3   The silencer, as its name implies, effectively reduces the exhaust noise to an acceptable level without having any adverse effects on engine performance. Do not tamper with or remove the baffles from within the silencer. Although a much louder exhaust note may give the impression of greater speed, this is rarely the case in practice. Apart from causing annoyance to others and getting motorcycling a bad name, tampering with the silencer usually results in a marked fall off in performance, often accompanied by a dramatic rise in the rate of petrol consumption.

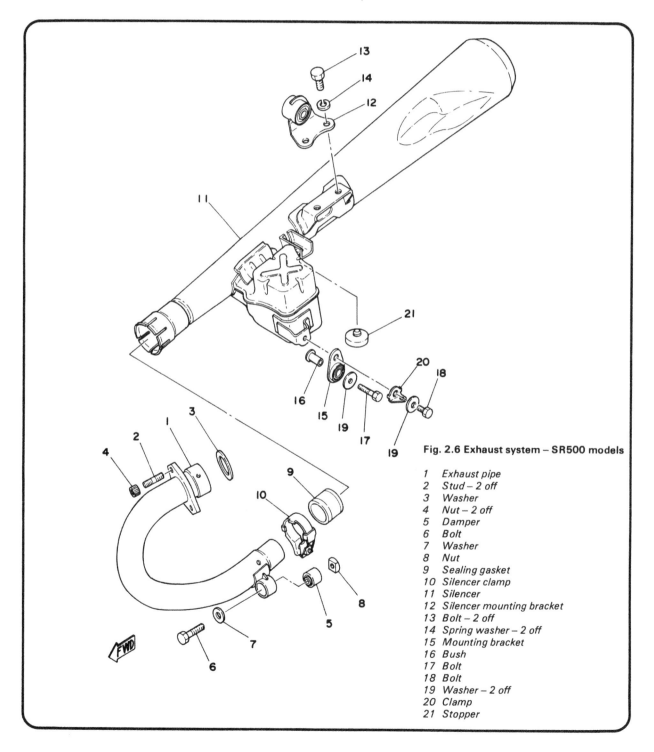

Fig. 2.6 Exhaust system – SR500 models

1   Exhaust pipe
2   Stud – 2 off
3   Washer
4   Nut – 2 off
5   Damper
6   Bolt
7   Washer
8   Nut
9   Sealing gasket
10 Silencer clamp
11 Silencer
12 Silencer mounting bracket
13 Bolt – 2 off
14 Spring washer – 2 off
15 Mounting bracket
16 Bush
17 Bolt
18 Bolt
19 Washer – 2 off
20 Clamp
21 Stopper

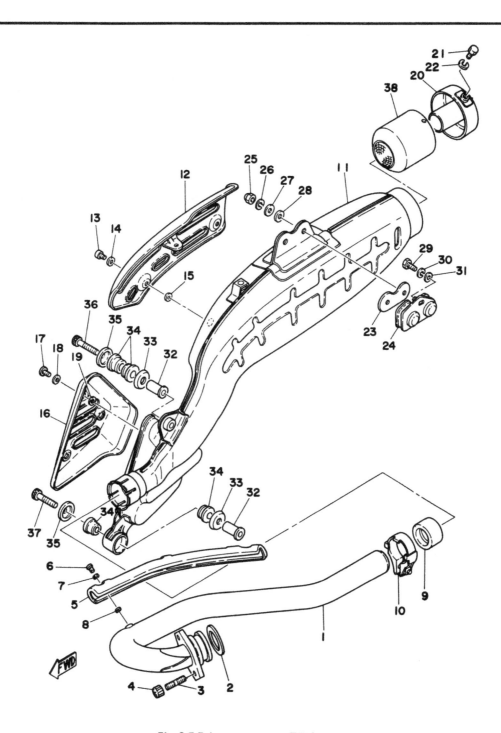

Fig. 2.7 Exhaust system –XT500 models

1	Exhaust pipe	14	Washer – 3 off	27	Plain washer – 2 off
2	Exhaust gasket	15	Washer – 3 off	28	Plain washer – 2 off
3	Stud – 2 off	16	Heat shield	29	Bolt
4	Knurled socket nut – 2 off	17	Screw – 3 off	30	Spring washer
5	Heat shield	18	Washer – 3 off	31	Plain washer
6	Screw – 2 off	19	Washer – 3 off	32	Collar
7	Washer – 2 off	20	Spark arrester cap	33	Dished washer
8	Washer – 2 off	21	Screw – 2 off	34	Rubber mounting – 2 off
9	Joint seal	22	Washer – 2 off	35	Dished plate
10	Clamp	23	Plate	36	Socket screw
11	Silencer	24	Mounting	37	Socket screw
12	Heat shield	25	Dome nut – 2 off	38	Spark arrester
13	Screw – 3 off	26	Spring washer – 2 off		

11 Engine lubrication

1   Engine lubrication for all the models in the Yamaha 500 single cylinder range is provided by a dry sump system. In this system the oil reservoir is contained in the interconnected frame top tube and front drum tube. The oil is fed to the engine via a feed pipe connected at one end to the bottom of the downtube and at the other end to the oil feed passage on the right-hand side of the crankcase. Oil is returned to the oil tank by a similar pipe connected to the engine behind the cylinder barrel.

2   Oil is transmitted by what is in effect two separate oil pumps of the trochoid type mounted on and driven by the same drive spindle. The outer pump is the main feed pump which is contained in its own casing. The inner pump, which is the scavenge (return) pump, is housed in the crankcase material, being enclosed by the main pump casing.

3   A full flow paper element oil filter is fitted to prevent continued circulation of any abrasive matter in the oil. In addition to this, two gauze oil strainers are incorporated to prevent larger particles of foreign matter from being passed through each oil pump.

4   The need to change the engine oil at regular intervals and at the same time to clean out or renew the filters in the system need not be overstressed. If the machine is used only for short journeys, especially in the lower temperature range, it is prudent to make the oil changes much more frequent. Low running temperatures encourage condensation within the engine and if the lubricating oil becomes contaminated with water, it will emulsify, leading to severe corrosion of many of the working parts.

5   The oil pump is unlikely to give trouble unless particles of metal find their way into the rotor assembly and cause scoring. This will usually only occur if some engine component has broken up during service and underlines the necessity to completely strip and clean the engine if any such unfortunate incident occurs. Impurities in the oil will normally be trapped by one or other of the filters, before they can cause any harm.

12 Oil pump: removal, dismantling, examination and replacement

1   In order to gain access to the oil pump the engine oil must be drained and the primary drive cover and clutch removed. Refer to Chapter 1, Sections 5.1–3, 12 and 13 for relevant details of the removal procedure. The procedure for oil pump removal is described in Chapter 1, Section 14 and that for reassembly in Section 36. It should be stressed that the oil pump assembly should not be removed and dismantled unless really essential. It will give good service during the normal service life of the machine without attention and is likely to give trouble only if metallic particles or other foreign bodies contaminate the oil and score the pump rotors.

2   After oil pump removal has been accomplished continue as follows in dismantling. Separate the main pump outer cover from the rotor housing. The cover is located on two close fitting dowels which should be pushed out to free the cover. Lift out the inner rotor and displace the drive pin which is a light sliding fit in the spindle. Lift out the outer rotor from the main pump casing. Withdraw the pump spindle, together with the scavenge pump inner rotor and drive pin. The scavenge pump inner rotor and drive pin can now be removed from the pump shaft.

3   Clean all the pump components thoroughly in petrol and then dry them before carrying out examination. Check each set of rotors and the housing in which they run for scoring or flaking. If damage to the rotors or the main pump casing is evident, the component in question should be renewed. Always renew each rotor set as a pair, not individually. The scavenge pump outer rotor runs directly in the crankcase. In the unlikely event of the housing becoming scored, the crankcase half must be renewed.

4   Refit the main pump rotors into the casing and check the clearance between the outer rotor and housing and between the inner rotor and outer rotor tips. Repeat this check with the scavenge pump rotors, after placing them in the crankcase and locating the inner rotor with the drive spindle. If the clearances are greater than those given in the Specifications, the rotors must be renewed.Also given in the Specifications at the beginning of the Chapter are the rotor thickness and housing depth wear limits. Components which are excessively worn must be renewed.

5   Check carefully the condition of the two oil seals in the main pump casing. These seals are, of course, vital to the efficient function of the oil pump and therefore the lubrication system. If there is any doubt about the condition of the seals they **must** be renewed. Each seal may be driven out using a parallel shanked drift of suitable diameter. When driving in the new seal ensure that it remains square with the housing, to prevent distortion.

6   Reassemble the oil pump by reversing the dismantling procedure. Absolute cleanliness must be observed at all times during reassembly; even a small piece of grit can cause severe damage to the rotors. Take special care when inserting the pump spindle through either oil seal lip. The flat, milled on the end of the spindle has a tendency to crease the fine sealing lip of the seal, causing permanent damage. During reassembly of the pump components apply copious quantities of clean engine oil as each is fitted. Note the triangular punch mark on one face of each rotor. When reassembling the pump, the punch marks on the scavenge pump should be facing towards the housing in the crankcase. Similarly, the main pump rotors should be fitted with the marks visible when the rotors are in the main pump casing.

13 Oil filter and strainer gauze: removal and cleaning

1   Access to the oil filter element and the two strainers can be made only after the engine oil has been drained from the crankcase and the oil tank. Renewal of the filter element and cleaning of the strainers should be carried out at every second oil change; that is at 4000 mile (6400 km) intervals. Refer to the relevant sections of the Routine Maintenance Chapter for details.

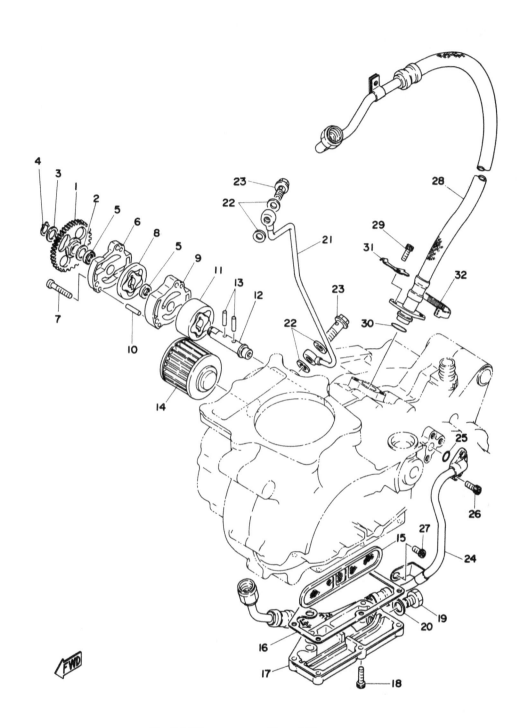

Fig. 2.8 Oil pump assembly and filter components

1	Driven pinion	12	Drive shaft	23	Banjo bolt – 2 off
2	Shim	13	Drive pin – 2 off	24	Oil feed pipe
3	Shim	14	Oil filter element	25	'O' ring
4	Circlip	15	Oil strainer screen	26	Screw – 2 off
5	Oil seal – 2 off	16	Gasket	27	Screw
6	Main pump cover	17	Sump cover	28	Oil return pipe
7	Screw – 3 off	18	Screw – 6 off	29	Screw – 2 off
8	Feed pump rotor set	19	Drain plug	30	'O' ring
9	Feed pump body	20	Sealing washer	31	Wiring clamp
10	Dowel– 2 off	21	Rocker feed pipe	32	Cable tie
11	Scavenge pump rotor set	22	Sealing washer – 4 off		

12.2a Separate the pump cover from the rotor housing (main pump)

12.2b Remove the inner rotor and drive pin and ...

12.2c ... push out the spindle to free the scavenge rotor

12.4a Check outer rotor/housing clearance and ...

12.4b ... inner rotor/outer rotor tip clearance

12.6a Punch marks on both pairs of rotor should ...

12.6b ... face outwards as shown

14 Fault diagnosis: fuel system and lubrication

Symptom	Cause	Remedy
Engine gradually fades and stops	Fuel starvation	Check vent hole in filler cap. Sediment in filter bowl or blocking float needle. Dismantle and clean.
Engine runs badly. Black smoke from exhaust	Carburettor flooding	Dismantle and clean carburettor. Check for punctured float or sticking float needle.
Engine lacks response and overheats	Weak mixture Air cleaner disconnected Modified silencer has upset carburation	Check for partial blockage in carburettor. Reconnect. Check hose for splits. Replace with original.
Engine loses power and gets noisy	Lubrication failure	Stop engine immediately and do not re-run until fault is located and remedied.

Chapter 3 Ignition system

Refer to Chapter 7 for information relating to the 1979 to 1983 models

Contents

Specifications

Flywheel generator	TT500C, D and E	XT500C, D and E
Type	Two-coil generator	Two-coil generator
Make	Nippon Denso	Nippon Denso
Output	6 volt	6 volt
Source coil resistance	2.214 ohms ± 10% at 20°C (68°F)	2.13 ohms ± 10% at 20°C (68°F)

Flywheel generator	SR 500
Type	Multi-coil alternator/CDI unit
Make	Nippon Denso
Output	12 volt
Pulser coil resistance:	
High speed	16 ohms ± 30% at 20°C (68°F)
Low speed	87 ohms ± 30% at 20°C (68°F)

Ignition timing	TT500C, D and E	XT500C, D and E	SR 500
Retarded	7°BTDC	7° BTDC	NA
Full advance	27° ± 3° BTDC	27° ± 3° BTDC	26.5°
Range	20° ± 3°	20° ± 3°	NA
Advancer system	centrifugal	centrifugal	electrical
Advance starts	2250 ± 150 rpm (TT500D and E 2100 $^{+300}_{-0}$ rpm)	2250 ± 150 rpm (XT500D and E 2100 $^{+300}_{-0}$ rpm)	1950 rpm
Advance finishes	3000 rpm ± 200	3000 rpm ± 200	6000 rpm

Contact breaker (except SR500)

Gap	0.3 – 0.4 mm (0.012 – 0.016 in)
Spring pressure	700 – 900g (24.7 – 31.7 oz)

Condenser (except SR500)

Capacity	0.22 microfarads ± 10%
Insulation resistance	3 Megohms or more

Ignition coil	TT and XT models	SR 500
Make	Nippon Denso	Nippon Denso
Type	029700 – 3900	029700 – 468
Primary winding resistance	0.75 ohms ± 10% at 20°C	0.98 ohms ± 20% at 20°C
Secondary winding resistance	5.75 K ohms ± 20% at 20°C	12.0 K ohms ± 20% at 20°C

Spark plug

Gap	0.7 – 0.8 mm (0.028 – 0.031 in)	
Make	NGK	Champion
Type	BP-7ES	N-7Y

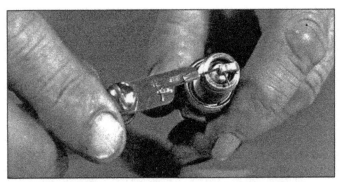
Electrode gap check - use a wire type gauge for best results

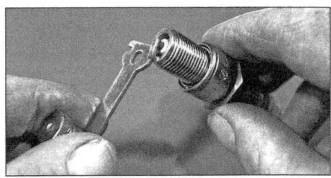
Electrode gap adjustment - bend the side electrode using the correct tool

Normal condition - A brown, tan or grey firing end indicates that the engine is in good condition and that the plug type is correct

Ash deposits - Light brown deposits encrusted on the electrodes and insulator, leading to misfire and hesitation. Caused by excessive amounts of oil in the combustion chamber or poor quality fuel/oil

Carbon fouling - Dry, black sooty deposits leading to misfire and weak spark. Caused by an over-rich fuel/air mixture, faulty choke operation or blocked air filter

Oil fouling - Wet oily deposits leading to misfire and weak spark. Caused by oil leakage past piston rings or valve guides (4-stroke engine), or excess lubricant (2-stroke engine)

Overheating - A blistered white insulator and glazed electrodes. Caused by ignition system fault, incorrect fuel, or cooling system fault

Worn plug - Worn electrodes will cause poor starting in damp or cold weather and will also waste fuel

1   General description

With the exception of the SR500 model, all the machines covered in this manual use a traditional ignition system comprising a flywheel generator, a single contact breaker assembly and an ignition secondary coil. The flywheel generator is mounted on the extreme left-hand end of the crankshaft and incorporates a permanent magnet rotor, rotating around a fixed stator to which is attached a charging/lighting coil and the ignition source coil. The charging/lighting coil and the ignition source coil are completely separate components which are not interconnected in any way. Where no battery or lights are used, the lighting coil is redundant, and although it is left in position, it is not interconnected with any circuit. The contact breaker camshaft is driven by a shaft/pinion assembly which itself is driven by a pinion on the right-hand end of the crankshaft.

The spark necessary to ignite the petrol/air mixture in the combustion chamber is derived from the flywheel generator ignition source coil. Current is produced immediately the engine is rotated and this is fed to a separate ignition coil mounted under the petrol tank, via the contact breaker assembly. When the contact breaker points separate, the primary circuit is broken, causing a high tension voltage to be developed in the secondary windings of the ignition coil. The voltage jumps across the sparking plug gap, to create the spark that ignites the mixture.

Unlike many other machines, the ignition system is not dependent on the battery. Because the machine is designed with off-road use in mind, the ignition system will function as normal when the battery has been removed from the machine, in conjunction with the lighting equipment.

SR500 models are fitted with a three phase alternator and utilise a CDI (capacitor discharge ignition) system. In this system the contact breaker is dispensed with and the ignition timing is controlled electronically by a magnetic pulser and pick-up in the alternator and a solid-state CDI unit which is mounted on the frame. By eliminating the mechanical contact breaker, frequent attention to the ignition system, and the inevitable alterations in ignition timing due to wear of moving parts, is eliminated.

2.3 Special extractor required for generator rotor removal

2   Ignition source coil: checking – TT and XT models

1   If the machine will not start and there is no evidence of a spark at the sparking plug, a check should first be made to ensure there is no fault at either the contact breaker assembly

or in the ignition coil itself. Refer to Sections 4 and 5 of this Chapter.

2   If the above checks have failed to show any faults, the output from the generator itself should be suspected. Failure to provide the correct output may be due to a number of faults which include a burnt out coil, short-circuit, open circuit or bad earthing due to loose coil mounting screws.

3   Failure in the circuit of the coil may be found by testing for the correct resistance, using an ohmmeter or a multi-meter set to the resistance scale. Carry out the test as follows. Track the flywheel generator leads up to the block type connector and separate the two halves of the connector. Connect one lead of the ohmmeter to the black or black/white wire from the generator and the other lead to a suitable earth point on the engine. The reading should be within the range given for each model as follows:

TT500	2·214 ohms ± 10% at 20°C (68°F)
XT500	2·130 ohms ±10% at 20°C (68°F)

If the resistance indicated is outside this range, the flywheel generator rotor must be removed to gain access for further investigation.

3   Remove the cover screws and lift off the cover. Select top gear and apply the rear brake firmly. This will prevent crankshaft rotation as the rotor centre nut is removed. The rotor is a tapered fit on the crankshaft end, located by a Woodruff key. Because of the method of attachment, the rotor will require pulling from position. To this end the rotor centre is provided with an internal thread, to take the Yamaha service tool No 90890–01189. This tool **must** be used to remove the flywheel rotor. There is insufficient room at the periphery to enable alternative types of puller, such as a two or three-legged sprocket puller to be fitted. No attempt should be made to lever the flywheel rotor from position. This will only result in damaged components. Screw the puller into the rotor until it is fully home, and then tighten down the puller centre screw. If the rotor is reluctant to move, do not continue tightening down the screw. A smart tap on the screw head with a hammer should break the joint between the tapers.

4   With the rotor removed, the ignition source coil may be examined. Check for signs of damage to the coil and for a broken main lead (black). External damage to the coil or lead will usually be self-evident. If no obvious signs of damage exist it is probable that the coil is faulty internally, and will have to be renewed. Check the coil screws for tightness, because a loose coil will give rise to a poor earth and consequent low performance.

3   Contact breaker: adjustments – TT and XT models

1   To gain access to the contact breaker assembly, it is necessary to detach the cover secured on the primary drive cover by two screws.

2   Rotate the engine slowly by means of the kickstarter until the points are in the fully-open position. Examine the faces of the contacts; if they are pitted or burnt, it will be necessary to remove them for further attention, as described in Section 5 of this Chapter.

3   Adjustment is carried out by slackening the two screws which retain the fixed contact breaker point and inserting a screwdriver in the adjusting slot provided. Turn in the appropriate direction until the gap is within the range 0·3 to 0·4 mm (0.012 to 0.016 inch) and then retighten the two retaining screws. It is imperative that the points are open FULLY whilst this adjustment is made, or a false reading will result.

4   Before replacing the cover and gasket, place a slight smear of grease on the contact breaker cam and a few drops of thin oil on the felt wick which lubricates the surface of the cam. Do not over lubricate the wick, because excess oil may find its way onto the points faces, causing ignition malfunction.

3.3a Check the points gap in the **fully open** position

3.3b Loosen these two screws to adjust the gap

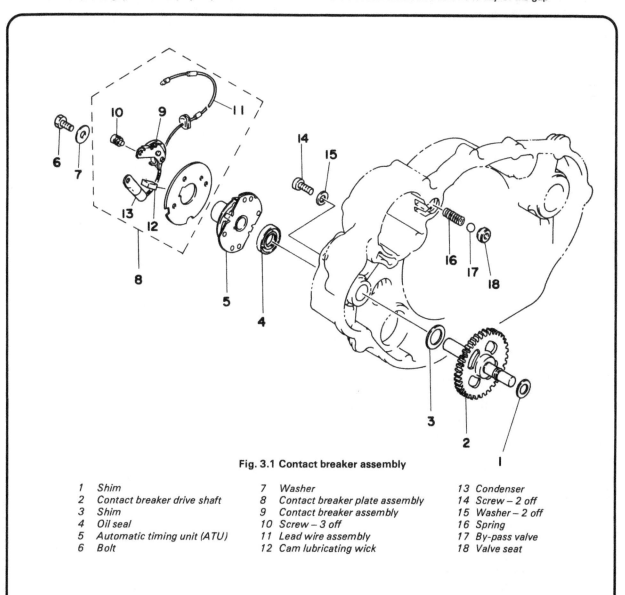

Fig. 3.1 Contact breaker assembly

1 Shim	7 Washer	13 Condenser
2 Contact breaker drive shaft	8 Contact breaker plate assembly	14 Screw – 2 off
3 Shim	9 Contact breaker assembly	15 Washer – 2 off
4 Oil seal	10 Screw – 3 off	16 Spring
5 Automatic timing unit (ATU)	11 Lead wire assembly	17 By-pass valve
6 Bolt	12 Cam lubricating wick	18 Valve seat

4   Contact breaker points: removal, renovation and replacement – XT and TT models

1   If the contact breaker points are burned, pitted or badly worn, they should be removed for dressing. If it is necessary to remove a substantial amount of material before the faces can be restored, the points should be replaced without question.
2   To remove the contact breaker points, detach the circlip which secures the moving contact to the pin on which it pivots. Remove the nut and bolt that secures the flexible lead wire to the end of the contact return spring, noting the arrangement of the insulating washers so that they are replaced in their correct order during reassembly. Lift the moving contact off the pivot, away from the assembly.
3   The fixed contact is removed by unscrewing the two screws which retain the contact to the contact breaker baseplate.
4   The points should be dressed with an oilstone or fine emery cloth. Keep them absolutely square throughout the dressing operation, otherwise they will make angular contact on reassembly and rapidly burn away.
5   Replace the contacts by reversing the dismantling procedure, making sure that the insulating washers are replaced in the correct order. It is advantageous to apply a thin smear of grease to the pivot pin, prior to replacement of the moving contact arm.
6   Check and, if necessary, re-adjust the contact breaker gap when the points are fully opened.

5   Condenser: removal and replacement – XT and TT models

1   A condenser is included in the contact breaker circuitry to prevent arcing across the contact breaker points as they separate. The condenser is connected in parallel with the points and if a fault develops, ignition failure is liable to occur.
2   If the engine proves difficult to start, or misfiring occurs, it is possible that the condenser is at fault. To check, separate the contact breaker points by hand when the ignition is switched on. If a spark occurs across the points and they have a blackened and burnt appearance, the condenser can be regarded as unserviceable.
3   It is not possible to check the condenser without the appropriate test equipment. In view of the low cost involved, it is preferable to fit a new replacement and observe the effect on engine performance.
4   The condenser is mounted on the contact breaker backplate by means of a metal clamp soldered to its body. Since this fitting provides the earth connection of the condenser, it follows that the clip should make good contact and be tightened fully.

4.2 Disconnect the LT lead and remove the circlip to separate points

6   Alternator: checking the output – SR500 models

1   The ac (alternating current) generator fitted to SR500 models not only fulfils the purpose of supplying current to the charging system and lights but also provides current to power the CDI unit and a pulser signal that controls the ignition timing. If the performance of the output coils in question is suspect, they may be tested with the alternator in place on the machine.
2   Follow the main alternator lead up from the engine to the block connector.Separate the connector so that the ends of the individual wires are visible. Using an ohmmeter or multi-meter set to the resistance scale check the following wiring combinations for resistance and note the readings.

Pulser coil resistance

High speed	*White/Red to Black*
	16 ohms $\pm$ *30% at 20°C (68°F)*
Low speed	*White/Green to Black*
	87 ohms $\pm$ *30% at 20°C (68°F)*

Charge coil resistance

High speed	*Red to Brown*
	3.5 - 6.5 ohm at 20°C (68°F)
	Red to Black
	334 ohm $\pm$ *30% at 20°C (68°F)*
Low speed	*Brown to Black*
	329 ohms $\pm$ *30% at 20°C (68°F)*

If any one reading is outside the given range, the coil that gives the incorrect resistance is damaged in some way. Because the alternator stator is an integrated unit, the complete assembly must be renewed.
3   If the readings taken are found to be correct, but the ignition system as a whole is malfunctioning, the ignition coil and the CDI unit should be checked to eliminate the faulty component.

7   CDI unit: testing – SR500 models

1   The solid-state CDI unit is mounted to the rear of the battery carrier behind the left-hand side panel. The unit is sealed for life and if any malfunction occurs, it must be renewed. Because no test data is available for the CDI unit, and because in any event testing requires very specialised equipment, malfunction of this component should be determined initially by a process of elimination. If the alternator is found to be satisfactory after testing as described in the preceding Section, and the ignition coil is in good order (see Section 8), it may be assumed that the CDI unit is faulty. Before consigning the unit to the rubbish bin, it is advised that the machine is returned to a Yamaha Service Agent for further tests to be carried out.

5.4 The condenser is held by a single screw

8 Ignition coil: checking

XT and TT models

1   The components most likely to fail in the ignition system are the condenser and the ignition coil since contact breaker faults should be obvious on close examination. Replacement of the existing condenser will show whether the condenser is at fault, leaving by the process of elimination the ignition coil.

2   The ignition coil can best be checked using a multi-meter set to the resistance position. Disconnect the black lead from the coil at the snap connector and remove the suppressor cap from the HT lead. Earth the negative lead from the multi-meter on the metal coil body. Make two tests; the first with the multi-meter positive lead connected to the black coil lead, and the second test with the positive lead connected to the HT lead. The results should be as follows if the ignition coil is in good condition.

Primary winding	*0·75 ohms ± 10% at 20°C (68°F)*
Secondary winding	*5·7 K ohms ± 20% at 20°C (68°F)*

SR500 models

3   A similar test to that described above should be carried out after disconnecting the block connector connecting the coil leads and removing the suppressor cap from the HT lead. To check the primary winding connect the meter's negative lead to the black coil wire and its positive lead to the orange coil wire. For the secondary winding check leave the meter's negative lead connected to the black wire, but connect its positive lead to the HT lead. If the coil is in good working order, the results should be within the range as follows:

Primary winding	*0·98 ohms ± 20% at 20°C (68°F)*
Secondary winding	*12·0 K ohms ± 20% at 20°C (68°F)*

All models

4   When taking resistance readings it should be noted that variations in the results may be encountered if the ambient temperature differs greatly from that given. Some allowance must be made in that as the temperature reduces so does the resistance. The converse also applies.

5   The ignition coil is a sealed unit and it is not possible to effect a satisfactory repair in the event of failure. A new coil must be fitted.

8.5 Ignition coil is located below the fuel tank

9 Ignition timing: checking and adjusting

1   Remove the alternator (flywheel generator) cover from the left-hand side of the machine so that access may be made to the rotor. It can be seen that the periphery of the rotor is inscribed with three lines; one marked T which indicates TDC, one marked F which is the ignition firing point, and one unmarked line which is the point at which full advance should occur. A small pointed projection on the casing wall above the rotor is the timing index mark, used as a reference point.

XT and TT models

2   Ignition timing should be checked and if necessary adjusted only after the contact breaker points have been cleaned and reset as described in Section 3 of this Chapter.

3   Apply a spanner to the flywheel generator nut and rotate the engine in a forward (anti-clockwise) direction. If the ignition timing is correct, the contact breaker points will be on the verge of opening when the scribed line marked F on the rotor aligns with the index pointer projecting from the casing. To determine the position at which the contact breaker points separate a battery and bulb arrangement, or a multimeter set on the resistance scale, can be used. Connect one probe lead to earth and the other to the contact breaker terminal or spring blade. As the engine is turned, the bulb will go out (or the meter needle will indicate infinity) as the contact faces separate. See Fig. 3.2 for details of the test lamp or multimeter connections.

4   If the timing is incorrect, reposition the rotor F mark so that it aligns with the index pointer. Slacken the two screws that pass into the contact breaker housing and secure the circular back plate. Turn the back plate in the required direction until the points are on the verge of opening and then tighten the screws. Rotate the engine backwards about 45° and then forwards slowly, until the F mark aligns again to recheck the setting. It is necessary to rotate the engine first backwards and then forwards so that all the slack in the contact breaker drive train is taken up.

5   Provided that the contact breaker is in good condition and care is taken, manual adjustment of the ignition timing should be acceptably accurate. If possible, however, the timing should be checked using a stroboscope lamp because not only can the accuracy of the timing be checked with the engine running but the correct performance of the ATU can be verified. The timing light should be connected to the low tension or high tension side of the ignition as instructed by the manufacturers of the light. Start the engine and illuminate the index pointer and rotor periphery with the stroboscope. Refer to the Specifications at the beginning of this Chapter to determine the advance characteristics for the engine being timed. From tick-over to the specified advance commencement rpm the F mark should align with the index pointer. As the advance rpm is reached and then exceeded, the F mark should be seen to move away from the index pointer. Between 2500 and 3200 rpm the advance should finish and the full advance scribed line (unmarked) on the rotor should be aligned with the index pointer.

6   If when increasing the engine speed from the commencement of advance the timing marks are seen to move erratically, or if the advance range has altered appreciably, the ATU should be inspected for wear or malfunctioning as described in the following Section.

SR500 models

7   The SR500 utilises an ignition system where the normal contact breaker unit is dispensed with; the ignition firing point being controlled entirely by electrical means. Because no points are used, checking the ignition timing can only be carried out when the engine is running, using a stroboscopic lamp. It should be noted that because no mechanical components are used, and therefore wear is eliminated, the ignition timing should remain accurate almost indefinitely during normal usage.

8   Connect the strobe lamp into the system as recommended by the manufacturer of the lamp. Start the engine and aim the lamp at the index pointer and the rotor periphery. Between tick-over and 1950 rpm the F mark on the rotor should be aligned with the index pointer. Slowly increase the engine speed past 1950 rpm. The F mark should move away from the index pointer. Ignition advance should be progressive until 6000 rpm when full advance should be reached and the advance scribed line on the rotor should be in alignment with the index pointer. Because of the very high full advance engine speed the advance performance should be checked as quickly as possible. Free running the engine at higher rpm settings for more than a few moments may cause damage.

9   If the ignition timing is found to be incorrect, adjustment may be carried out within certain limits. The alternator stator on which the ignition magnetic pick-up and coils are secured is retained in the casing by three screws passing through elongated holes in the stator plate periphery. With the engine at rest, a screwdriver may be passed through the aperture in the flywheel (rotor) face and the three screws loosened. Rotate the stator a small amount in the required direction and then tighten the screws. Rotate the stator plate clockwise to advance the ignition and anti-clockwise to retard the ignition. After tighten-ing the screws fully, start the engine and re-check the timing. If further adjustment is needed, repeat the procedure. It can be seen that re-adjusting the ignition timing must be by the trial and error method. Moving the stator plate should be done a little at a time.

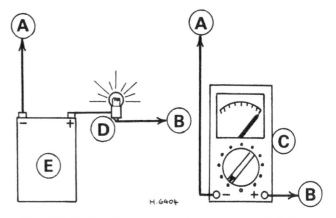

Fig. 3.2 Static timing test using battery and bulb or multimeter

A   Connect to moving contact terminal
B   Connect to earth (ground)
C   Multimeter set on resistance scale
D   Bulb
E   Torch battery

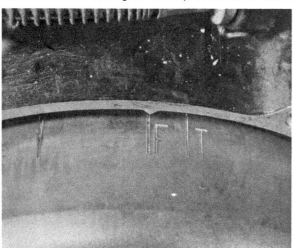

9.2a F mark on rotor relates to firing point (retarded)

9.2b Unmarked line relates to full advance position

9.2c Timing index mark may be used for static timing only

9.4 Ignition timing adjustment screws

10 Automatic timing unit: examination

1 The automatic timing unit rarely requires attention although it is advisable to examine it periodically.

2 To obtain access to the unit remove the inspection cover and the contact breaker back plate complete with contact breaker. The ATU centre bolt should be removed before the back plate. Before removal, the back plate and end cover should be marked so that the back plate can be replaced in exactly the same position. This will ensure the ignition timing is not altered.

3 Pull the ATU from position, noting the drive pin with which it locates and is driven. The unit comprises balance weights which move outwards against spring tension as the centrifugal force increases. The balance weights must move freely on their pivots, which should be lubricated. The tension springs must also be in good condition.

4 Check the surface of the contact breaker cam for pitting or obvious signs of wear. Damage to the cam cannot be rectified; the complete ATU must be renewed.

5 When replacing the ATU, check that the drive pin engages with the recess in the rear of the centre boss. Because there is a single recess only, the ATU cannot inadvertently be replaced in the incorrect position and so alter the timing marks in relation to the crankshaft.

11 Sparking plug: checking and resetting the gap

1 All models are fitted with an NGK BP–7ES or Champion N–7Y sparking plug as standard, gapped to within the range 0·7 – 0·8 mm (0·028 – 0·031 in). Operating conditions may indicate a change in sparking plug grade; the type recommended by the manufacturer gives the best, all round service.

2 Check the gap of the plug points during every 2000 mile service. To reset the gap, bend the outer electrode to bring it closer to the centre electrode and check that a 0·7 – 0·8 mm (0·028 – 0·031 in) feeler gauge can be inserted. Never bend the central electrode or the insulator will crack, causing engine damage if the particles fall in whilst the engine is running.

3 With some experience, the condition of the sparking plug electrodes and insulator can be used as a reliable guide to engine operating conditions. See accompanying illustrations.

4 Beware of overtightening the sparking plug, otherwise there is risk of stripping the threads from the aluminium alloy cylinder head. The plug should be sufficiently tight to sit firmly on its copper sealing washer, and no more. Use a spanner which is a good fit to prevent the spanner from slipping and breaking the insulator.

5 If the threads in the cylinder head strip as a result of over tightening the sparking plug, it is possible to reclaim the head by the use of a Helicoil thread insert. This is a cheap and convenient method of replacing the threads; most motorcycle dealers operate a service of this kind.

6 Make sure the plug insulating cap is a good fit and has its rubber seal. It should also be kept clean to prevent tracking. This cap contains the suppressor that eliminates both radio and TV interference.

12 Fault diagnosis

Sympton	Cause	Remedy
Engine will not start	Faulty ignition switch	Operate switch several times in case contacts are dirty. A faulty switch must be renewed.
	Short circuit in wiring	Check whether fuse is intact. Eliminate fault before switching on again.
Engine misfires	Faulty condenser in ignition circuit (XT and TT models)	Replace condenser and retest.
	Fouled spark plug	Replace plug and have original cleaned.
	Poor spark due to generator failure	Check output from generator.
Engine lacks power and overheats	Retarded ignition timing	Check timing Check contact breaker (XT and TT models) Check whether auto-advance mechanism has jammed (XT and TT models) Check electronic advance performance with strobe light. (SR 500 models)
Engine 'fades' when under load	Pre-ignition	Check grade of plugs fitted; use recommended grades only. Check lubrication system.

Chapter 4 Frame and forks

Refer to Chapter 7 for information relating to the 1979 to 1983 models

Contents

Specifications

	TT500C and XT500C	TT500D and XT500D	TT500E and XT500E	SR500
Front forks				
Type		Hydraulically damped, telescopic		
Oil capacity (per leg)	217 cc (7.3/6.1 US/Imp fl oz)	223 cc (7.5/6.3 US/Imp fl oz)	223 cc (7.5/6.3 US/Imp fl oz)	182 cc (6.1/5.1 US/Imp fl oz)
Oil type		SAE 20 fork oil or SAE 10W/30 engine oil		SAE 10
Fork travel	195 mm (7.68 in)	195 mm (7.68 in)	195 mm (7.68 in)	150 mm (5.9 in)
Spring free length	491 mm (19.35 in)	515 mm (20.30 in)	445 mm (17.54 in)	450 mm (17.71 in)
Rear suspension				
Type		Swinging arm, controlled by two gas/oil suspension units		
Wheel travel	144 mm (5.67 in)	159 mm (6.26 in)	159 mm (6.26 in)	110 mm (4.33 in)
Spring length	249 mm (9.80 in)	269 mm (10.6 in)	269 mm (10.6 in)	216 mm (8.52 in)
Swinging arm free play		1 mm (0.039 in) max		

1 General description

All the models within the Yamaha 500cc single cylinder range utilise a similar full-cradle frame of welded tubular construction. In this type of frame the engine/gearbox unit does not comprise part of the frame as a stressed member. Because the engine has a dry sump lubrication system the oil must be contained remotely on some part of the frame. To save the extra weight and space required for a special oil reservoir, the oil is stored in the interconnected frame top and downtubes.

Rear suspension is of the swinging arm type, using oil filled suspension units to provide the necessary damping action. The units are adjustable so that the spring ratings can be effectively changed within certain limits to match the load carried.

The front forks are of the conventional telescopic type, having internal, oil-filled dampers. The fork springs are contained within the fork stanchions and each fork leg can be detached from the machine as a complete unit, without dismantling the steering head assembly.

2 Front fork legs: removal from the frame

1 It is unlikely that the front forks will have to be removed from the frame as a complete unit, unless the steering head bearings require attention or if the machine suffers frontal damage due to an accident. Removal of the fork legs should be carried out as described in this Section, and then, if necessary, further dismantling may take place as described in the following Section.

2 Commence operations by removing the front wheel, following the procedure given in Chapter 5 Section 4. Where no centre stand is fitted, the machine must be supported securely on blocks with the front wheel clear of the ground. In any event, sufficient clearance must be given to allow each fork leg to be pulled downwards, clear of the fork yokes.

3 The SR500 model is fitted with a disc brake and the caliper unit must be detached from the left-hand fork leg at this stage. Loosen and remove the two bolts which pass through the body of the caliper and into the fork leg. The caliper unit can then be

lifted clear of the disc and secured to some part of the frame or engine using a length of wire. Disconnection of the hose from the caliper is not required but care must be taken not to apply the front brake or the piston may be expelled, causing loss of hydraulic fluid and the resulting need for subsequent bleeding of the brake system.

4   On SR500 models unscrew the two bolts which pass into each fork leg and secure the mudguard in place. The mudguard on all other models is retained by four bolts passing upwards into the fork lower yoke. Because of this the mudguard need only be removed if removal of the complete forks, including the yokes, is anticipated.

5   Slacken the pinch bolts which pass into the fork yokes and which clamp the fork legs in place. Similarly, on XT500 models, loosen the screws which clamp the headlamp mounting brackets to the fork stanchions. The individual fork legs may be pulled downwards out of the upper and lower yokes and away from the machine. On XT500 models the headlamp should be supported as the second fork leg is withdrawn, and then allowed to hang from the wiring loom.

2.5a Slacken the upper and lower pinch bolts

2.5b Loosen the headlamp bracket screws (XT 500 only)

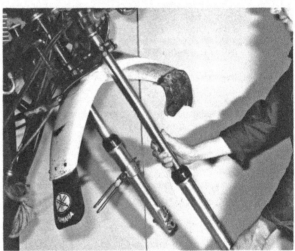
2.5c Withdraw each fork leg individually

3   Fork yokes: removal

1   If the fork yokes require removal either for renewal or to enable access to be gained to the steering head bearings, the fork legs should first be removed as described in the preceding Section. Having accomplished that, continue dismantling as follows.

2   Remove the headlamp glass/reflector unit from the headlamp (where fitted) so that access can be made to the wiring connections. The reflector unit is secured by two screws passing into the shell at the 4 o'clock and 8 o'clock positions (SR500) or a single screw in the 4 o'clock position (XT500). Disconnect the headlamp main bulb socket and the pilot lamp bulb connection (where fitted) and lift the freed reflector unit away. Further disconnection of electrical leads should be carried out only after the battery has been isolated by removing one or more leads from the battery terminals. If this precaution is not taken inadvertent short circuiting may take place due to bared wiring ends earthing against cycle parts. Disconnect the leads at the block connectors within the headlamp shell as required, until the leads can be pulled through the grommet in the rear of the shell. On XT models the headlamp shell complete with the mounting brackets may be lifted away. On SR500 models the

headlamp bracket is a one-piece unit. The shell and bracket can be removed after unscrewing the two lower mounting bolts.

3   On SR500 models remove the single bolt which passes into the lower yoke and secures the hydraulic brake hose junction piece. Detach the brake master cylinder/handlebar lever unit from the handlebars. The assembly is secured by a clamp held by two screws. The master cylinder and caliper unit can now be removed from the machine, still interconnected by the hydraulic brake hose. If this method of removal is adopted, and the hydraulic hose is not disconnected, bleeding of the brake system on reassembly will not be required. Do not allow the contents of the master cylinder reservoir to spill. Hydraulic brake fluid is a very good paint stripper.

4   Place a thick layer of rag over the front of the petrol tank to protect the paint, and then detach the handlebars, complete with controls. The bars are secured by two half-clamps held by two screws each. Rest the handlebars on the tank.

5   Disconnect the speedometer and the tachometer drive cables at the instrument heads by unscrewing the knurled securing rings, and then detach the wiring leads from the switches and instruments at the block connectors. The instruments share a common mounting bracket secured by two bolts. Once these have been removed, the instruments may be lifted clear.

6   Loosen the clamp bolt located at the rear of the upper yoke, and from the top of the yoke remove the large chrome bolt together with the washer. From the underside, tap the upper yoke upwards until it frees the steering column. Support the weight of the lower yoke and, using a 'C' spanner, remove the steering head bearing adjuster ring. If a 'C' spanner is not available, a soft brass drift and hammer may be used to slacken the nut.

7   Remove the dust excluder and outer race (cone) once the adjuster nut has been detached. The bottom yoke, complete with steering column, can now be lowered from position. Make provision to catch the ball bearings as they are released; only the lower bearings will drop free since the upper bearings will most probably remain seated in the cup race retaining them.

4   Front forks: dismantling

1   It is advisable to dismantle each fork leg separately, using an identical procedure. There is less chance of unwittingly exchanging parts if this approach is adopted. Commence by draining each fork leg of damping oil; there is a drain plug in each lower leg above and to the rear of the wheel spindle housing.

2   Clamp the fork lower leg in a vice fitted with soft jaws, or wrap a length of rubber inner tube around the leg to prevent damage. Unscrew the socket screw, recessed into the housing which carries the front wheel spindle. Move the fork leg so that the fork stanchion is held by the vice; do not overtighten or the tube may distort. On SR500 models prise the rubber cap from the top of the stanchion. Using a socket key, unscrew the recessed plug, to allow access to the fork spring. On all other models a chromed bolt is fitted in place of the plug. Remove the sleeve (where fitted), the spring seat and the fork spring. Note that some fork springs are fitted that have closer pitches at one end. The spring must be refitted in the same position on subsequent reassembly.

3   Prise the dust excluder from position and slide it up the fork upper tube. On XT500E and TT500E models a long rubber gaiter is fitted in place of the dust excluder. Before this can be

removed the screw clip securing each end must be slackened. The fork stanchion, complete with the damper rod assembly, can now be withdrawn from the fork lower leg.

4   As can be seen from the accompanying diagrams the type of damper rod assembly used depends upon the type of machine to which the forks are fitted. On XT and TT models displace the circlip from the lower end of the stanchion after pulling off the damper rod seat to aid access. Remove the piston, collar, valve block and rebound spring from the damper rod and then invert the stanchion and allow the rod to slide out. No further dismantling of the fork leg is possible. On SR500 models pull the seat from the end of the damper rod, invert the stanchion and push out the damper rod towards the top. The piston ring need not be removed from the piston integral with the end of the damper rod.

5   The oil seal fitted to the top of the lower leg should be removed only if it is to be renewed, because damage will almost certainly be inflicted when it is prised from position. The seal is retained by a spring clip, under which is fitted a large washer.

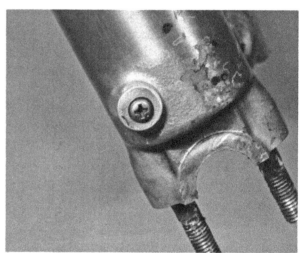

4.1 A drain plug is fitted in each fork lower leg

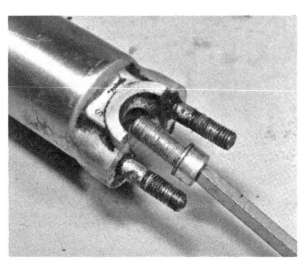

4.2a Remove the damper rod securing socket screw

4.2b Unscrew the stanchion top bolt and ...

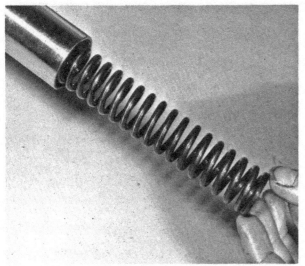

4.2c ... withdraw the fork spring

4.3a Prise off the dust excluder and ...

4.3b ... separate the stanchion from the lower leg

4.4a Pull off the damper rod seat (hydraulic stop piece)

4.4b Displace the stanchion circlip to allow removal of ...

4.4c ... the piston, collar, valve block and spring

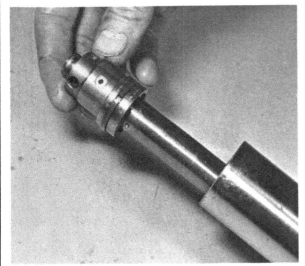

4.4d Remove the damper rod from the upper end of the stanchion

Fig. 4.1 Front forks – XT500 models

1 Front fork assembly
2 LH lower leg
3 RH lower leg
4 Oil seal – 2 off
5 Backing washer – 2 off
6 Circlip – 2 off
7 Circlip – 2 off
8 Gaiter clip – 2 off
9 Gaiter clip
10 Gaiter – 2 off
11 Gaiter clip – 2 off
12 Screw – 2 off
13 Damper piston – 2 off
14 Damper rod assembly
15 Stanchion – 2 off
16 Fork spring – 2 off
17 Spring seat – 2 off
18 Spacer – 2 off
19 'O' ring – 2 off
20 Cap bolt – 2 off
21 Steering stem
22 Lower yoke
23 Pinch bolt
24 Spring washer
25 Nut
26 Pinch bolt – 2 off
27 Pinch bolt – 2 off
28 Plain washer – 5 off
29 Clamp
30 Plain washer
31 Cable guide
32 Spring washer – 4 off
33 Nut – 3 off
34 Spindle clamp
35 Plain washer – 2 off
36 Nut – 2 off
37 Sealing washer
38 Socket screw
39 Sealing washer – 2 off
40 Drain plug – 2 off
41 LH Headlamp bracket
42 RH Headlamp bracket
43 Inner damper rubber – 2 off
44 Outer damper rubber – 2 off
45 Screw – 4 off
46 Spring washer – 4 off
47 Plain washer – 4 off
48 Cable guide
49 Cable guide
50 Bracket
51 Bolt
52 Spring washer
53 Plain washer
54 Reflector bracket – 2 off
55 Reflector damper
56 Reflector

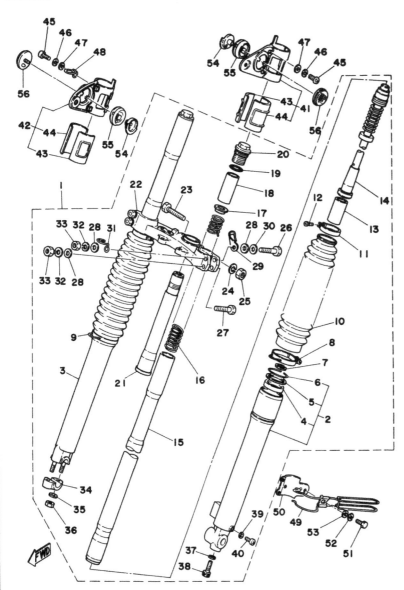

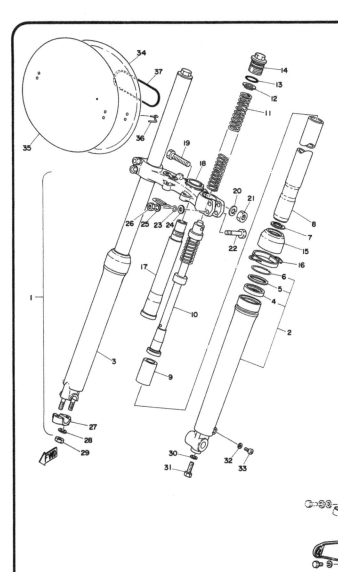

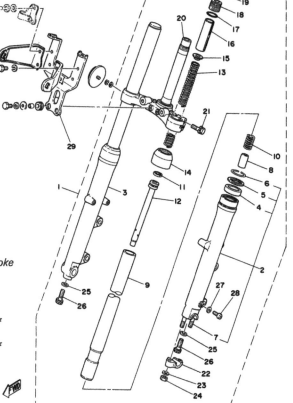

Fig. 4.2 Front fork assembly – TT500 models

1	Front fork assembly	20	Spring washer
2	LH lower leg	21	Nut
3	RH lower leg	22	Pinch bolt – 4 off
4	Oil seal	23	Cable guide
5	Backing plate	24	Plain washer – 4 off
6	Circlip – 2 off	25	Spring washer – 4 off
7	Circlip – 2 off	26	Nut – 4 off
8	Stanchion – 2 off	27	Spindle clamp
9	Damper piston – 2 off	28	Plain washer – 2 off
10	Damper rod – 2 off	29	Nut – 2 off
11	Fork spring – 2 off	30	Sealing washer
12	Spring upper seat – 2 off	31	Bolt
13	'O' ring – 2 off	32	Sealing washer – 2 off
14	Cap bolt – 2 off	33	Drain plug – 2 off
15	Dust excluder – 2 off	34	Number plate
16	Gaiter clip – 2 off	35	Number plate
17	Steering stem	36	Wire stay – 2 off
18	Lower yoke	37	'O' ring
19	Pinch bolt		

Fig. 4.3 Front fork assembly – SR500 models

1	Front fork assembly	16	Spacer – 2 off
2	LH lower leg	17	O-ring – 2 off
3	RH lower leg	18	Cap bolt – 2 off
4	Oil seal – 2 off	19	Cap – 2 off
5	Backing plate – 2 off	20	Steering stem/lower yoke
6	Circlip – 2 off	21	Pinch bolt – 4 off
7	Stud – 2 off	22	Spindle clamp
8	Damper rod seat – 2 off	23	Spring washer – 2 off
9	Stanchion – 2 off	24	Nut – 2 off
10	Rebound spring – 2 off	25	Sealing washer – 2 off
11	Piston ring – 2 off	26	Socket screw – 2 off
12	Damper rod – 2 off	27	Sealing washer – 2 off
13	Spring – 2 off	28	Drain screw – 2 off
14	Dust excluder – 2 off	29	Headlamp bracket
15	Spring seat – 2 off		

5 Front forks: examination and renovation

1   The front forks are not provided with bushes. The fork legs slide directly against the hard chrome surface of the stanchions and if wear occurs, either the stanchion and/or the lower fork leg will have to be renewed. Wear normally occurs only after a very considerable mileage has been covered and can be detected by a juddering sensation when the front brake is applied. A slack steering head assembly will give the same effect, so this should always be checked first and adjusted if necessary, before condemning the forks.

2   Wear is often visually apparent in the form of scuffing or break-through of the chrome surface of the fork stanchions. If evidence of damage of this nature is apparent, the stanchion in question must be removed. In extreme cases the fork lower leg will have to be renewed as well.

3   If damping action is lost, the piston ring around the damper piston should be inspected for wear or loss of tension. Unfortunately, the piston ring is not available as a separate item and if wear or damage has occurred, the complete rod must be renewed. Check that the small holes in the damper tubes are not blocked. If extreme blockage has occurred but no substantial improvement is shown when the forks are reassembled and refilled, renew the complete damper assembly.

4   It is rarely possible to straighten forks which have been badly damaged in an accident, especially if the correct jigs are not available. It is always best to err on the side of safety and fit new replacements, especially since there is no easy means of checking to what extent forks have been overstressed. The fork stanchions can be checked for straightness by rolling them on a flat surface. Any misalignment will immediately be obvious.

5   The fork springs may show signs of compression after lengthy service, in which case they can be replaced to advantage. The correct spring lengths are given in the Specifications section of this Chapter.

6 Steering head bearings: examination and renovation

1   Before commencing reassembly of the forks, examine the steering head races. The ball bearing tracks of the respective cup and cone bearings should be polished and free from indentations and cracks. If signs of wear or damage are evident, the cups and cones must be replaced. They are a tight push fit and should be drifted out of position.

2   Note that when each head race bearing has its full complement of ball bearings they are not packed tightly and that suf-ficient room is left to accommodate one extra ball. This spacing is essential because if the bearings are packed tightly against each other, they will skid rather than roll, thus greatly accelerating the rate of wear. The number and size of the balls in each bearing differ and care should be taken not to transpose one type of ball from one race to another. The upper bearing has 22 balls of $\frac{3}{16}$ inch diameter and the lower race 19 balls of $\frac{1}{4}$ inch diameter.

7 Front forks: replacement

1   Replace the front forks by reversing the dismantling procedure. When fitting the damper components to the XT and TT 500 forks, ensure that they are in the correct relative position and sequence. Refer to the photographs which accompany the text.

2   The fork legs must be filled with the correct quantity and grade of damping fluid before the top bolts (or plugs) are refitted. This may be done either during reassembly of the fork legs or after they have been installed in the yokes. In either case, check that the drain plugs are fitted and tightened, before refitting.

3   On XT and TT500 models insert the fork legs in the yokes so that the alignment groove near the upper end of each stanchion is flush with the top face of the upper yoke. Doing this will ensure that each fork leg extends the correct amount from the yokes. On SR500 models the top edge of the stanchion should be flush with the upper yoke top surface.

4   The fork yoke pinch bolts should be tightened only after the front wheel has been re-installed as described in Chapter 5 Section 4 and after the steering head bearings have been adjusted. Adjustment of the bearings is described in the next Section.

8 Steering head bearings: adjustment

1   The adjustment of the steering head bearings should be checked at regular intervals as a part of normal routine maintenance. Additionally, re-adjustment of the bearings should be carried out whenever the fork yokes have been removed.

2   If the bearing adjustment is too slack judder will occur. There should be no play at the head races when the handlebars are pulled and pushed, with the front brake fully applied. Overtight head races are equally undesirable. It is possible to unwittingly apply a pressure of several tons on the head bearings by overtightening, even though the handlebars appear to turn quite

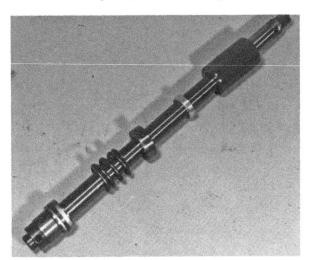

5.3a Damper rod assembly: component parts

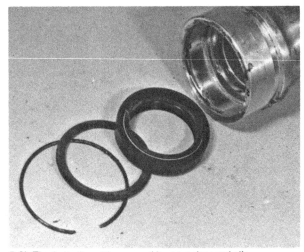

5.3b The fork oil seal is secured by a washer and clip

freely. Overtight bearings will cause the machine to roll at low speeds and give generally imprecise handling. Adjustment is correct if there is no play in the bearings and the handlebars swing to full lock either side when the machine is on the centre stand with the front wheel clear of the ground. Only a light tap on each end should cause the handlebars to swing.

3   Adjustment is effected by tightening or loosening the adjuster ring using a 'C' spanner. Note that the pinch bolt at the rear of the upper yoke, and the large crown bolt which passes into the top of the yoke must be loosened, before adjustment is attempted.

9   Steering head lock

1   The steering lock on XT models is fitted to the left-hand side of the steering head lug. When the key is turned and the steering is on full-lock, a tongue projecting from the lock engages with a recess in the steering column. The lock is retained by a single rivet which serves also as a pivot for the lock cover. To remove the lock, the rivet must be drilled out; this operation requires considerable care to ensure that the rivet hole is not enlarged.

2   On SR500 models the steering lock is incorporated in the ignition switch and works on a principle similar to that

described above. If the lock malfunctions, the complete switch must be renewed.

10   Frame: examination and renovation

1   The frame is unlikely to require attention unless accident damage has occurred. In some cases, replacement of the frame is the only satisfactory course of action if it is badly out of alignment. Only a few frame repair specialists have the jigs and mandrels necessary for testing the frame to the required standard of accuracy and even then there is no easy means of assessing to what extent the frame may have been overstressed.

2   After the machine has covered a considerable mileage, it is advisable to examine the frame closely for signs of cracking or splitting at the welded joints. Rust corrosion can also cause weakness at these joints. Minor damage can be repaired by welding or brazing, depending on the extent and nature of the damage.

3   Remember that a frame which is out of alignment will cause handling problems and may even promote 'speed wobbles'. If misalignment is suspected, as the result of an accident, it will be necessary to strip the machine completely so that the frame can be checked and, if necessary, renewed.

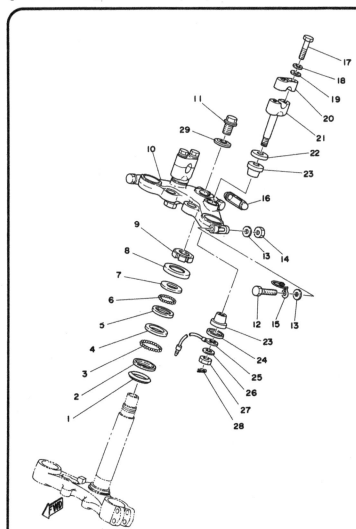

Fig. 4.4 Fork yokes and steering head bearings

1   Dust seal
2   Cup
3   Ball – 19 off
4   Cone
5   Cup
6   Ball – 22 off
7   Cone
8   Cover
9   Adjuster ring
10   Upper yoke
11   Bolt
12   Bolt – 3 off
13   Washer – 6 off
14   Nut – 3 off
15   Cable holder
16   Cap
17   Bolt – 4 off
18   Spring washer – 4 off
19   Plain washer – 4 off
20   Handlebar clamp – 2 off
21   Handlebar holder – 2 off
22   Washer – 2 off
23   Bush – 4 off
24   Washer – 2 off
25   Earth lead
26   Washer – 2 off
27   Nut – 2 off
28   Spring clip – 2 off
29   Washer

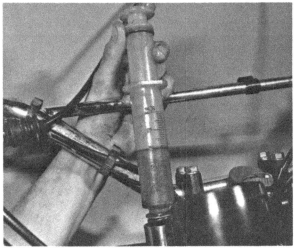

7.2a Refill the forks with the correct quantity of damping fluid

7.2b Do not omit the O ring on the stanchion bolt

11 Swinging arm rear fork: dismantling, examination and renovation

1   The rear fork of the frame assembly pivots on two needle roller bearings, one of which is pressed into each end of the swinging arm crossmember. A central spacer running the length of the crossmember acts as an inner race for both bearings, and the complete unit is supported by a pivot shaft passing through the frame lugs.

2   Worn swinging arm pivot bearings will give imprecise handling with a tendency for the rear end of the machine to twitch or hop. The play can be detected by placing the machine on its centre stand and with the rear wheel clear of the ground, pulling and pushing on the fork ends in a horizontal direction. Any play will be greatly magnified by the leverage effect. It is comparatively easy to renovate the swinging arm pivot and bushes when wear necessitates attention to the rear suspension.

3   To remove the swinging arm fork, first place a stout wooden box or other support under the frame so that the rear of the machine is raised off the ground. On SR500 models the machine should be placed on the centre stand. Remove the rear wheel as described in Chapter 5 Section 13.

4   On SR500E models, withdraw the split pin from the caliper support pivot bolt and then remove the bolt and nut to free the caliper unit. Move the caliper unit forward, still connected by the hydraulic brake hose, and attach it to some part of the frame so that it will be clear of further dismantling.

5   Removal of the chainguard, and where fitted, the lower chain protector, is not strictly necessary, although removal will give more room for manoeuvring. This also applies to the chain tension arm fitted to some models.

6   Detach the lower end of each rear suspension unit from its mounting point on the swinging arm. On TT models, where the units are mounted on studs, it may be necessary to slacken the upper mounting nuts to allow sufficient lateral movement for the units to be pulled off the lower studs. XT500 and SR500 have bolts supporting the units and therefore this problem will not arise.

7   Remove the nut (and where fitted, the washer) from the left-hand end of the swinging arm pivot shaft. Support the weight of the swinging arm fork and withdraw the shaft. Where the shaft is corroded into place it will have to be tapped out, using a suitable drift. Take care not to damage the threads, on the shaft end. With the shaft removed, the swinging arm may be withdrawn from the rear of the machine.

8   Pull off the chain protection collar from the left-hand end of the swinging arm cross-member and then remove the swinging arm bearing covers. These covers should not be interchanged as the internal components differ from one side to another. In addition to the rubber seal and thrust bearing common to both sides of the cross-member, the left-hand cover is fitted with a thrust washer. If this is fitted inadvertently to the right-hand cover, the swinging arm will be off-centre. Push the central long spacer from position in the cross-member so that access to the bearings is made.

9   Do not attempt to remove the bearings until they have been cleaned and examined, and then only if renewal is dictated by their condition. The bearings must be driven from position and this will almost certainly damage the cages. If the bearings are worn or if pitting has occurred due to neglect of lubrication, they must be renewed. Use a long drift passed through from the opposite side of the cross-member to drift each bearing out in turn. If possible, the new bearings should be pressed into position. If this is not possible they may be driven into place using a tubular drift whose outer diameter is similar to that of the bearing cage. Take care not to allow the bearing to tilt when being fitted because this will distort the cage. Check the condition of the centre spacer. The ends of the spacer from the inner race of the bearings. If pitting or flaking is evident, the spacer should be renewed.

10 Reassemble and refit the swinging arm fork by reversing the dismantling procedure. Grease the two bearings thoroughly before assembly and then apply a grease gun to the nipple provided, after installation of the swinging arm. Continue pumping grease into the cross-member until it can be seen to be forced out of the ends of the bearing covers. Wipe off the excess lubricant. Remember that the single thrust washer must be fitted to the left-hand bearing cover, between the thrust bearing and cover, if correct centralisation of the swinging arm is to be maintained.

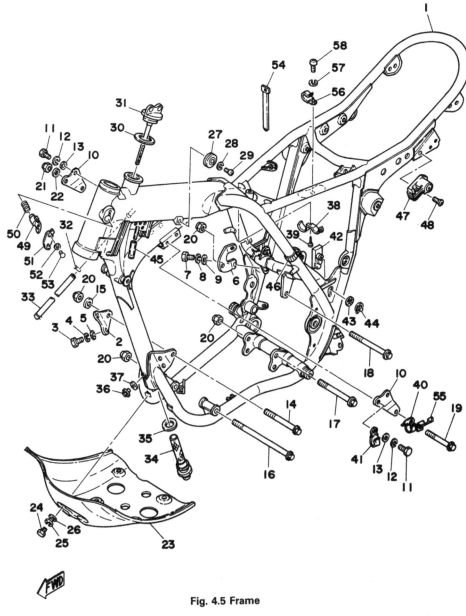

Fig. 4.5 Frame

1	Frame
2	Front mounting bracket
3	Bolt – 2 off
4	Spring washer – 2 off
5	Plain washer – 2 off
6	Rear mounting plate
7	Bolt – 2 off
8	Spring washer – 2 off
9	Plain washer – 2 off
10	Head steady bracket – 2 off
11	Bolt – 4 off
12	Spring washer – 4 off
13	Plain washer – 4 off
14	Bolt
15	Washer
16	Bolt
17	Bolt
18	Bolt
19	Bolt
20	Self locking nut – 4 off
21	Self locking nut
22	Conical spring washer
23	Sump guard
24	Bolt – 6 off
25	Spring washer – 6 off
26	Plain washer – 6 off
27	Tank rubber – 2 off
28	Spring washer – 2 off
29	Screw – 2 off
30	Sealing ring
31	Filler cap/dipstick
32	Spring clip – 2 off
33	Vent tube
34	Feed union/filter
35	Sealing washer
36	Drain plug
37	Sealing washer
38	Oil pipe clamp
39	Screw
40	Cable guide
41	Cable clip
42	Chain protector
43	Wave washer
44	Circlip
45	Hose
46	Cable tie
47	Helmet lock
48	Screw
49	Steering lock
50	Conical spring
51	Lock cover
52	Washer
53	Rivet
54	Cable tie
55	Cable guide
56	Cable guide
57	Spring washer
58	Screw

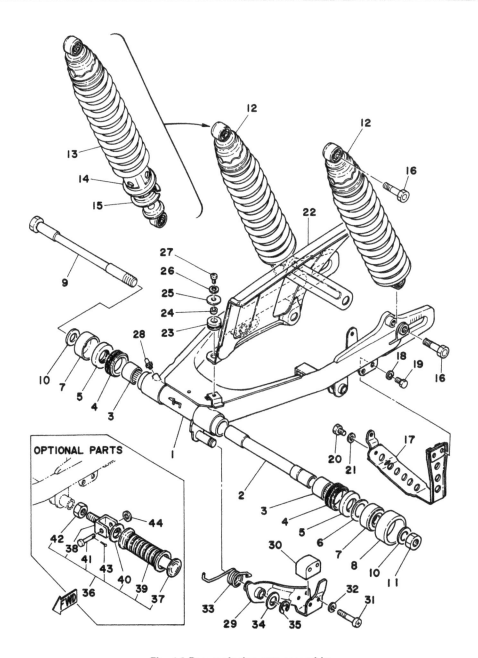

Fig. 4.6 Rear swinging arm assembly

1	Swinging arm fork	
2	Central spacer (bush)	
3	Needle roller bearing – 2 off	
4	Seal – 2 off	
5	Thrust bearing – 2 off	
6	Shim	
7	Bearing cover – 2 off	
8	Chain protector	
9	Swinging arm pivot	
10	Washer – 2 off	
11	Nut	
12	Rear suspension unit – 2 off	
13	Suspension spring – 2 off	
14	Spring guide – 2 off	
15	Collar – 2 off	
16	Bolt – 4 off	
17	Lower chain guard	
18	Spring washer – 2 off	
19	Bolt – 2 off	
20	Bolt	
21	Spring washer	
22	Chain guard	
23	Grommet – 3 off	
24	Collar – 3 off	
25	Plain washer – 3 off	
26	Spring washer – 3 off	
27	Screw – 3 off	
28	Grease nipple	
29	Chain tensioner arm	
30	Tensioner block	
31	Screw – 2 off	
32	Spring washer – 2 off	
33	Tension spring	
34	Plain washer	
35	'E' clip	
36	Rear footrest assembly – 2 off – optional	
37	Footrest bar	
38	Footrest stock	
39	Rubber	
40	Plain washer	
41	Clevis pin	
42	Nut	
43	Split pin	
44	Washer	

11.6 Detach the lower end of each suspension unit

11.7a Remove the nut and ...

11.7b ... withdraw the swinging arm fork pivot

11.7c The fork may be lifted out to the rear

11.8a Remove the chain protector (left-hand side only) and ...

11.8b ... pull off the bearing cover from each side

11.8c Note the rubber seal and ...

11.8d ... the thrust bearing fitted to each cover

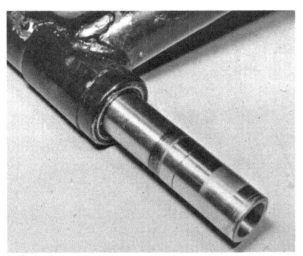

11.8e Withdraw the long central spacer

11.9 The bearings may be drifted out

12 Rear suspension units: examination

1   Rear suspension units of the five position hydraulically damped type are fitted to all models. They can be adjusted to give five different spring settings without need for their removal from the frame, so that the suspension characteristics can be matched to suit the riding conditions. As a general guide the softest setting is recommended for road use only, when no pillion passenger is carried. The hardest setting should be used when a heavy load is carried, and during high-speed riding either on or off road. The intermediate positions may be used as conditions dictate.

2   Each unit has two peg holes immediately above the adjusting notches to facilitate adjustment. Either a 'C' spanner or a metal rod can be used to turn the adjusters.

3   There is no means of draining the units or topping up, because the dampers are built as a sealed unit. If the damping fails or if the units commence to leak, the complete damper assembly must be renewed. This applies equally if the damper rod has become bent.

4   The damping efficiency of the units can best be judged after removal of the springs. The compression springs can be removed by detaching the damper units from the machine and holding each one upright on the workbench whilst the top is pressed downwards in opposition to the spring pressure. This will permit a second person to remove the split collets from the top so that the springs can be withdrawn over the upper end of the unit. On XT and TT models the damper units are mounted upside down and should therefore be inverted for dismantling. The damping characteristics differ depending upon the design use of the machine, and this should be borne in mind when assessing the damping performance.

5   On SR models, heavy damping should be apparent when the damper is compressed, and little resistance should be felt during damper extension. On all other models the characteristics are the converse.

6   If the damper springs have become weakened or are of differing lengths they should be renewed as a matched pair. The spring free lengths and the maximum allowable shortening of the springs are given in the Specifications at the beginning of this Chapter.

7   The dampers fitted to XT and TT models contain nitrogen gas under high pressure and for this reason some care must be taken when handling the units. The safety instructions issued by the manufacturers are similar to those associated with aerosol

cans and are as follows:

    a) Do not subject the dampers to a naked flame of high heat source because this may expand the gas beyond the limits of the container, thereby causing an explosion.

    b) Do not damage or deform the damper cylinder in any way other than that recommended in the following paragraph.

    c) Before a damper unit is discarded, the gas pressure must be released. This action should be carried out to prevent accidental damage to other people at a later date. To release the pressure a 2-3 mm (0.08-0.12 in) hole should be drilled in the wall of the damper cylinder 10-15 mm (0.40-0.60 in) from the cylinder end nearest the damper eye.

When carrying out this operation **eye protection** in the form of goggles or a visor **must be** worn to prevent damage from escaping gas, or from swarf particles created by the drill. This operation is quite straightforward provided that sensible precautions are taken.

8   In the interest of good roadholding, it is essential that both units are set to the same load setting and that when replacements are made, both units are treated in an identical manner so that they stay matched.

13 Centre stand: examination – SR models only

1   The centre stand is secured to two large lugs, welded to the frame lower tubes, by two shouldered bolts which serve as pivot points. The stand is returned by a single extension spring.

2   To prevent seizure of the pivot bolts and rapid wear of the surfaces, the bolts should be removed at intervals, for cleaning and relubrication. A heavy graphited or waterproof grease is recommended for this application.

3   Check the condition of the return spring, bearing in mind that if the spring breaks in service the stand will drop down and may cause an accident.

14 Prop stand: examination

1   A prop stand is fitted so that the machine can be parked without difficulty. It bolts to a lug on the underside of the frame and when retracted, lies parallel to the frame, well out of the way. An extension spring ensures the prop stand is returned to the fully retracted position when the weight of the machine is taken off it and it is pushed backwards with the foot.

2   Periodically check that the pivot bolt is secure and that the extension spring is in good condition and not over-stretched. An accident is more or less inevitable if the stand extends whilst the machine is on the move.

15 Footrests: examination and renovation

1   The front footrests fitted to XT and TT models are of the all metal competition type. Each footrest is supported on a splined stub shaft projecting from the frame, where it is secured by a pinch bolt and retained axially by a bolt and large washer. Each rider's footrest on SR500 models is supported on two studs passing through a frame lug and isolated from vibration by a rubber damper on each stud. Domed nuts are used to hold the footrest on the studs. On all models the footrests are hinged and spring loaded so that they may fold upwards if an object is struck.

2   Pillion footrests are provided as standard on SR500 models, and as an optional extra on XT500 models. In common with the front footrests these too are spring loaded.

3   If the footrests become damaged in an accident, it may be possible to straighten them after removal from the frame. The area around the deformed portion should be heated to a dull red before any attempt is made to bend the footrest back into shape. If required, the hinged portion of the footrest may be

separated from the bracket after removing the clevis pin or roll pin. If there is evidence of failure of the metal either before or after straightening, it is advised that the damaged component is renewed. If a footrest breaks in service, loss of machine control is almost inevitable.

16 Rear brake pedal: examination and renovation

1   The rear brake pedal has a splined centre and is attached to a pivot that passes through a lug on the lower right hand side of the frame. It is retained by a pinch bolt and can be varied in position by the way in which it is aligned with the splined pivot. A coil spring around the pivot ensures the pedal is returned to its normal operating position after braking.

2   If the pedal is bent or twisted in an accident, it should be removed from the splined pivot and clamped in a vice. Straighten the pedal using the method recommended for footrests in the preceding Section. The warning relating to footrest breakage applies equally to the brake pedal because it follows that failure is most likely to occur when the brake is applied firmly, which is when it is required most.

17 Dualseat: removal and replacement

1   The dualseat fitted to all models is secured by two bolts towards the rear which pass through or into frame lugs close to the seat base. The front of the seat is located by a tongue or slots engaging with projecting bolts.

2   Removal of the seat is straightforward; unscrew the bolts and pull the seat to the rear so that it disengages at the front. The seat can then be lifted away.

18 Speedometer and tachometer heads: removal and replacement – XT and SR500 models only

1   The speedometer and tachometer heads are independent instruments sharing a common mounting bracket attached to the fork upper yoke. They may be removed individually using the same procedure or together, still attached to the mounting bracket.

2   To remove one instrument first detach the drive cable by unscrewing the knurled coupling ring. Unscrew the two domed nuts from the base of the mounting bracket, below the instrument. Lift the instrument up so that access can be made to the bulb holders; these are a light push fit and may be withdrawn easily.

3   Removal of the instrument assembly complete with the mounting bracket can be made by detaching the warning lamp leads at the connectors provided and then removing the two mounting bolts.

4   Apart from defects in either the drive or the drive cable, a speedometer or tachometer which malfunctions is difficult to repair. Fit a replacement or alternatively entrust the repair to a competent instrument repair specialist.

5   Remember that a speedometer in correct working order is a statutory requirement in the UK. Apart from this legal necessity, reference to the odometer reading is the most satisfactory means of keeping pace with the maintenance schedules.

19 Speedometer and tachometer drive cables: examination and maintenance

1   It is advisable to detach the drive cable(s) from time to time in order to check whether they are lubricated adequately, and whether the outer coverings are damaged or compressed at any point along their run. Jerky or sluggish movements can often be traced to a damaged drive cable.

2   For greasing, withdraw the inner cable. After removing all the old grease, clean with a petrol-soaked rag and examine the cable for broken strands or other damage.

3   Regrease the cable with high melting point grease, taking care not to grease the last six inches at the point where the cable enters the instrument head. If this precaution is not observed, grease will work into the head and immobilise the instrument movement.

4   If any instrument head stops working suspect a broken drive cable unless the odometer readings continue. Inspection will show whether the inner cable has broken; if so, the inner cable alone can be replaced and re-inserted in the outer casing, after greasing. Never fit a new inner cable alone if the covering is damaged or compressed at any point along its run.

20 Speedometer and tachometer drives: location and examination

1   In the case of the disc front brake machines, the speedometer drive gearbox is fitted on the right-hand side of the front wheel hub. On drum front brake machines the gearbox is an integral part of the brake plate and is driven internally from the front hub. In both cases the drive rarely gives trouble provided it is kept properly lubricated. Lubrication should take place whenever the front wheel is removed for wheel bearing inspection or replacement.

2   The tachometer drive is taken from the overhead camshaft by means of skew-cut pinions and then by a flexible cable to the tachometer head via the cylinder head cover. It is unlikely that the internal drive will give trouble during the normal service life of the machine, particularly since it is fully enclosed and effectively lubricated.

21 Fault diagnosis: frame and forks

Symptom	Cause	Remedy
Machine veers either to the left or the right with hands off handlebars	Bent frame	Check and renew.
	Twisted forks	Check and replace.
	Wheels out of alignment	Check and re-align.
Machine rolls at low speed	Overtight steering head bearings	Slacken until adjustment is correct.
Machine judders when front brake is applied	Slack steering head bearings	Tighten until adjustment is correct.
	Worn fork bushes	Dismantle forks and renew bushes.
Machine pitches on uneven surfaces	Ineffective fork dampers	Check oil content.
	Ineffective rear suspension units	Check whether units still have damping action.
	Suspension too soft	Raise suspension unit adjustment one notch.
Fork action stiff	Fork legs out of alignment (twisted in yokes)	Slacken yoke clamps, and fork top bolt. Pump fork several times then retighten from bottom upwards.
Machine wanders. Steering imprecise. Rear wheel tends to hop	Worn swinging arm pivot	Dismantle and renew bearings and pivot shaft.

Chapter 5 Wheels, brakes and tyres

Refer to Chapter 7 for information relating to the 1979 to 1983 models

Contents

Specifications

	TT500C and D	TT500E	XT500C, D and E	SR500
Tyres				
Front	3.00 x 21	3.00 x 21	3.00 x 21	3.50S19
Rear	4.60 x 18	4.00 x 18	4.00 x 18	4.00S18

Tyre pressures				
Front:				
Off road	13 psi	13 psi	13 psi	—
*On road	—	—	18 psi	26 psi
Rear:				
Off road	16 psi	16 psi	16 psi	—
*On road	—	—	21 psi	28 psi

*When carrying a passenger or travelling at continuous high speed increase the front tyre pressure by 2–3 psi and the rear tyre pressure by 4–5 psi

	TT500 C, D and E	XT500C, D and E	SR500
Brakes			
Front	130 mm (5.12 in) SLS drum	160 mm (6.30 in) SLS drum	298 mm (11.73 in) disc
Rear	160 mm (6.30 in) SLS drum	150 mm (5.91 in) SLS drum	267 mm (10.51 in) disc or 150 mm (10.51 in) drum

Hydraulic fluid (SR500 only)

Type ....   DOT 3 (USA), SAE J1703 (UK)

1 General description

All models are fitted with an 18 inch diameter rear wheel and, with the exception of the SR500 which has a 19 inch front wheel, a 21 inch diameter wheel at the front. The XT500 and TT500 models utilise traditional wheels, comprised of aluminium alloy rims and hubs laced together with steel spokes.

The SR500 features seven spoke cast aluminium wheels.

As can be seen by reference to the Specifications at the beginning of the Chapter, the type of brake fitted to each wheel is dependent upon the use for which the machine in question is designed. The drum brake is a standard single leading shoe type, and the disc brake is hydraulically operated, utilising a floating, single-piston caliper.

2 Front wheel: examination and renovation – wire spoked wheels

1 Place the machine on the centre stand so that the front wheel is raised clear of the ground. Spin the wheel and check the rim alignment. Small irregularities can be corrected by tightening the spokes in the affected area although a certain amount of experience is necessary to prevent over-correction. Any flats in the wheel rim will be evident at the same time. These are more difficult to remove and in most cases it will be necessary to have the wheel rebuilt on a new rim. Apart from the effect on stability, a flat will expose the tyre bead and walls to greater risk of damage if the machine is run with a deformed wheel.

2 Check for loose and broken spokes. Tapping the spokes is the best guide to tension. A loose spoke will produce a quite different sound and should be tightened by turning the nipple in an anticlockwise direction. Always check for run out by spinning the wheel again. If the spokes have to be tightened by an excessive amount, it is advisable to remove the tyre and tube as detailed in Section 22 of this Chapter. This will enable the protruding ends of the spokes to be ground off, thus preventing them from chafing the inner tube and causing punctures.

3 Front wheel: examination and renovation (cast alloy wheels)

1 Carefully check the complete wheel for cracks and chipping, particularly at the spoke roots and the edge of the rim. As a general rule a damaged wheel must be renewed as cracks will cause stress points which may lead to sudden failure under heavy load. Small nicks may be radiused carefully with a fine file and emery paper (No. 600 – No. 1000) to relieve the stress. If there is any doubt as to the condition of a wheel, advice should be sought from a Yamaha repair specialist.

2 Each wheel is covered with a coating of lacquer, to prevent corrosion. If damage occurs to the wheel and the lacquer finish is penetrated, the bared aluminium alloy will soon start to corrode. A whitish grey oxide will form over the damaged area, which in itself is a protective coating. This deposit however, should be removed carefully as soon as possible and a new protective coating of lacquer applied.

3 Check the lateral run out at the rim by spinning the wheel and placing a fixed pointer close to the rim edge. If the maximum run out is greater than 2·0 mm (0·08 in), Yamaha recommend that the wheel be renewed. This is, however, a council of perfection; a run out somewhat greater than this can

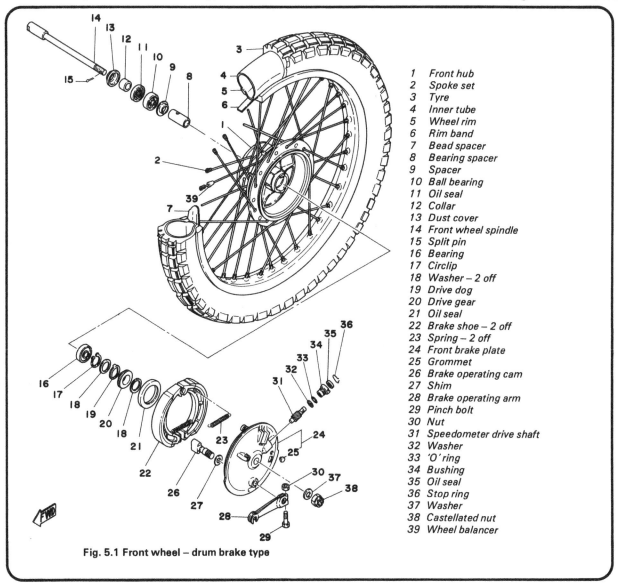

1 Front hub
2 Spoke set
3 Tyre
4 Inner tube
5 Wheel rim
6 Rim band
7 Bead spacer
8 Bearing spacer
9 Spacer
10 Ball bearing
11 Oil seal
12 Collar
13 Dust cover
14 Front wheel spindle
15 Split pin
16 Bearing
17 Circlip
18 Washer – 2 off
19 Drive dog
20 Drive gear
21 Oil seal
22 Brake shoe – 2 off
23 Spring – 2 off
24 Front brake plate
25 Grommet
26 Brake operating cam
27 Shim
28 Brake operating arm
29 Pinch bolt
30 Nut
31 Speedometer drive shaft
32 Washer
33 'O' ring
34 Bushing
35 Oil seal
36 Stop ring
37 Washer
38 Castellated nut
39 Wheel balancer

Fig. 5.1 Front wheel – drum brake type

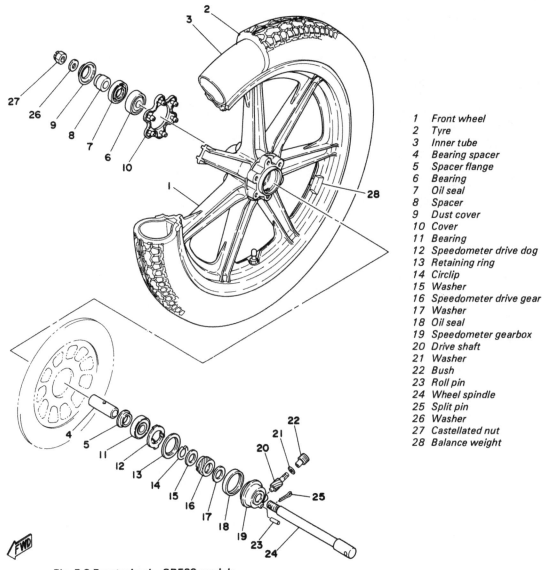

1   Front wheel
2   Tyre
3   Inner tube
4   Bearing spacer
5   Spacer flange
6   Bearing
7   Oil seal
8   Spacer
9   Dust cover
10 Cover
11 Bearing
12 Speedometer drive dog
13 Retaining ring
14 Circlip
15 Washer
16 Speedometer drive gear
17 Washer
18 Oil seal
19 Speedometer gearbox
20 Drive shaft
21 Washer
22 Bush
23 Roll pin
24 Wheel spindle
25 Split pin
26 Washer
27 Castellated nut
28 Balance weight

Fig. 5.2 Front wheel – SR500 models

probably be accommodated without noticeable effect on steering. No means is available for straightening a warped wheel without resorting to the expense of having the wheel skimmed on all faces. If warpage was caused by impact during an accident, the safest measure is to renew the wheel complete. Worn wheel bearings may cause rim run out. These should be renewed as described in Section 12 of this Chapter.

4   Front wheel: removal and replacement

1   Before the front wheel is removed, the machine must be supported securely on blocks so that the wheel is well clear of the ground. On SR500 models, place the machine on the centre stand first, to aid support. The blocks should be placed below the crankcase, taking care that they are so positioned that there is no danger of the machine rolling forwards.
2   Detach the speedometer drive cable (where fitted) at the wheel. On SR500 models the cable is retained by a knurled ring, and on all other models it is retained by an internal circlip. On drum brake models slacken the cable adjuster off to give sufficient slack in the cable to allow the nipple to be displaced

from the brake actuating arm. Pull the cable out of the abutment in the brake back plate.
3   Remove the split pin from the end of the wheel spindle and unscrew the castellated nut. After slackening the spindle clamp nuts, support the weight of the wheel and withdraw the spindle. The wheel can then be lowered from place and pulled forward from between the fork legs.
4   The front wheel may be refitted by reversing the dismantling procedure. On drum brake models ensure that the brake back plate anchor slot engages with the lug projecting from the left-hand fork leg. If this is not done, the back plate will rotate on the first application of the brake, causing the wheel to lock. When replacing the wheel on disc brake models ensure that the brake disc enters the caliper squarely and does not foul the brake pads. Insert the wheel spindle and then fit and tighten the nut fully. The spindle clamp should now be tightened down; tighten the front nut first to a torque wrench setting of 1·6 – 2.2 kgf m (12 – 16 lbf ft) and then tighten the rear nut to the same torque figure. A gap will be visible between the mating surfaces of the clamp and fork to the rear of the spindle. If the spindle clamp was removed, it must be refitted with the arrow mark provided facing forwards.

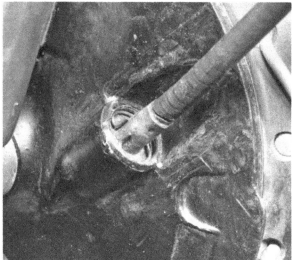
4.2a Prise out the circlip and ...

4.2b ... pull the speedometer cable from position

4.2c Disconnect the brake cable at the arm and ...

4.2d... withdraw the cable from the brake backplate

4.3a Remove the split pin and spindle nut

4.3b Slacken the clamp nuts fully and withdraw the spindle

4.4a Ensure that the fork leg lug engages with the slot in the plate

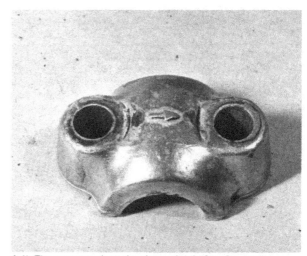

4.4b The arrow mark on the clamp should face forwards

5 Front drum brake: examination and renovation – XT and TT models only

1   With the front wheel removed, as described in the preceding section, the brake mechanism and backplate can be pulled free from the drum.
2   Examine the drum surface for signs of scoring or oil contamination. Both of these conditions will impair braking efficiency. Remove all traces of dust, preferably using a brass wire brush, taking care not to inhale any of it, as it is of an asbestos nature, and consequently dangerous. Remove oil or grease deposits, using a petrol soaked rag.
3   If deep scoring is evident, due to the linings having worn through to the shoe at some time, the drum must be skimmed on a lathe, or renewed. Whilst there are firms who will undertake to skim a drum whilst fitted to the wheel, it should be borne in mind that excessive skimming will change the radius of the drum in relation to the brake shoes, therefore reducing the friction area until extensive bedding in has taken place. Also full adjustment of the shoes may not be possible. If in doubt about this point, the advice of one of the specialist engineering firms who undertake this work should be sought.

4   If fork oil or grease from the wheel bearings has badly contaminated the linings, they should be renewed. There is no satisfactory way of degreasing the lining material, which in any case is relatively cheap to replace. It is a false economy to try to cut corners with brake components; the whole safety of both machine and rider being dependent on their condition.
5   The linings are bonded to the shoes, and the shoe must be renewed complete with the new linings. Removal is accomplished by folding the shoes together until the spring tension is relaxed, and then lifting the shoes and springs off the brake plate. Fitting new shoes is a direct reversal of the above procedure.
6   Before refitting existing shoes, roughen the lining surface sufficiently to break the glaze which will have formed in use.
7   The brake fulcrum pin (camshaft) should be removed for inspection. Before detaching the actuating arm so that the pin can be pressed out, mark the relative positions of each component with a centre punch, so that the arm can be refitted on the pin splines in the same position. The lubrication of the fulcrum pin is often neglected, leading to rapid wear of the cam faces and the bearing surfaces. In extreme cases, the bore of the brake back plate may wear, in which case both the plate and pin should be renewed. Regrease the pin before reassembly.

5.1 After wheel removal the drum brake may be removed as a unit

5.5 Check the linings for wear

6 Front disc brake: checking and renewing the pads – SR models only

1 To facilitate the checking of brake pad wear, the caliper is provided with an inspection window closed by a small cover. Prise the cover from position and inspect both pads. Each pad has a red wear limit line around its periphery. If either pad has worn down to or past the line, both pads in the set should be renewed.

2 Removal of the pads is straightforward, and does not require the hydraulic hose to be disconnected. Remove the two bolts that secure the caliper bracket to front fork leg, and lift the complete unit up off the brake disc. Remove the single bolt which secures the caliper unit to the caliper mounting bracket. Unscrew the crosshead screw from the inner face of the caliper, noting that the screw acts as a detent for the pads. Pull the support bracket from the main caliper and lift both pads from position. Note the various shims and their positions, before removal.

3 Fit new pads by reversing the dismantling sequence. If difficulty is encountered when fitting the caliper over the brake disc, due to the reduced distance between the new pads, use a wooden lever to push the pad on the piston side inwards.

4 In the interests of safety, always check the function of the brakes before taking the machine on the road.

7 Front disc brake: removing, renovating and replacing the caliper unit – SR models only

1 Before the caliper assembly can be removed from the fork leg upon which it is mounted, it is first necessary to drain off the hydraulic fluid. Disconnect the brake pipe at the union connection it makes with the caliper unit and allow the fluid to drain into a clean container. It is preferable to keep the front brake lever applied throughout this operation, to prevent the fluid from leaking out of the reservoir. A thick rubber band cut from a section of inner tube will suffice, if it is wrapped tightly around the lever and the handlebars.

2 Note that brake fluid is an extremely efficient paint stripper. Take care to keep it away from any paintwork on the machine or from any clear plastic, such as that sometimes used for instrument glasses.

3 When the fluid has drained off, remove the caliper mounting bolts, separate the two main caliper components and remove the pads as described in the preceding Section.

4 To displace the piston, apply a blast of compressed air to the brake fluid inlet. Take care to catch the piston as it emerges from the bore – if dropped or prised out with a screwdriver a piston may suffer irreparable damage. Before removing the piston, displace the dust seal which is retained by a circlip.

5 Remove the sleeve and protective boot upon which the caliper unit slides. If play has developed between the sleeve and the caliper, the former must be renewed. Check the condition of the boot, renewing it if necessary.

6 The parts removed should be cleaned thoroughly, using only brake fluid as the solvent. Petrol, oil or paraffin will cause the various seals to swell and degrade, and should not be used under any circumstances. When the various parts have been cleaned, they should be stored in polythene bags until reassembly, so that they are kept dust free.

7 Examine the piston for score marks or other imperfections. If it has any imperfections it must be renewed, otherwise air or hydraulic fluid leakage will occur, which will impair braking efficiency. With regard to the various seals, it is advisable to renew them all, irrespective of their appearance. It is a small price to pay against the risk of a sudden and complete front brake failure. It is standard Yamaha practice to renew the seals every two years, even if no braking problems have occurred.

8 Reassemble under clinically-clean conditions, by reversing the dismantling procedure. Apply a small quantity of graphite grease to the slider sleeve before fitting the boot. Reconnect the

hydraulic fluid pipe and make sure the union has been tightened fully. Before the brake can be used, the whole system must be bled of air, by following the procedure described in Section 11 of this Chapter.

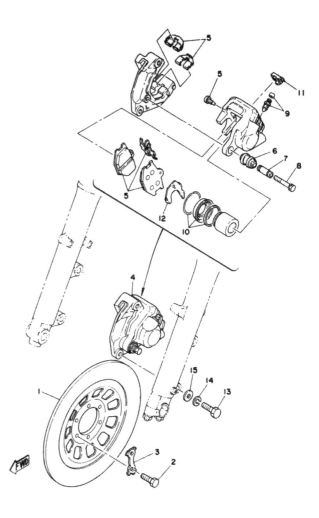

Fig. 5.3 Front brake caliper unit – SR500 models

1	Brake disc	9 Bleed nipple and cap
2	Bolt – 6 off	10 Seal kit
3	Locking plate – 3 off	11 Cap
4	Caliper assembly	12 Shim
5	Pad set	13 Bolt – 2 off
6	Bush boot	14 Spring washer – 2 off
7	Sleeve	15 Washer – 2 off
8	Bolt	

8 Removing and replacing the brake disc – SR models only

1 It is unlikely that the disc will require attention until a considerable mileage has been covered, unless premature scoring of the disc has taken place thereby reducing braking efficiency. To remove either the front disc, or where fitted the rear disc, the relevant wheel must first be removed. Refer to Section 4 for front wheel removal and Section 13 for rear wheel removal. The procedure for removal and examination of either disc is similar. The disc is bolted to the wheel by six bolts, which are secured in pairs by a common tab washer. Bend back the tab washers and remove the bolts, to free the disc.

Tyre changing sequence - tubed tyres

A Deflate tyre. After pushing tyre beads away from rim flanges push tyre bead into well of rim at point opposite valve. Insert tyre lever adjacent to valve and work bead over edge of rim.

Use two levers to work bead over edge of rim. Note use of rim protectors

B

C Remove inner tube from tyre

When first bead is clear, remove tyre as shown

D

E When fitting, partially inflate inner tube and insert in tyre

Work first bead over rim and feed valve through hole in rim. Partially screw on retaining nut to hold valve in place.

F

G Check that inner tube is positioned correctly and work second bead over rim using tyre levers. Start at a point opposite valve.

Work final area of bead over rim whilst pushing valve inwards to ensure that inner tube is not trapped

H

2   The brake disc can be checked for wear and for warpage whilst the front wheel is still in the machine. Using a micrometer, measure the thickness of the disc at the point of greatest wear. If the measurement is much less than the recommended service limit of 4·5 mm (0·18 in) the disc should be renewed. Check the warpage of the disc by setting up a suitable pointer close to the outer periphery of the disc and spinning the front wheel slowly. If the total warpage is more than 0·15 mm (0·006 in) the disc should be renewed. A warped disc, apart from reducing the braking efficiency, is likely to cause juddering during braking and will also cause the brake to bind when it is not in use.

9 Hydraulic brake hoses and pipes: examination – SR models only

1   An external brake hose and pipe is used to transmit the hydraulic pressure to the caliper unit when the front brake or rear brake is applied. The brake hose is of the flexible type, fitted with an armoured surround. It is capable of withstanding pressures up to 350 kg/cm². The brake pipe attached to it is made from double steel tubing, zinc plated to give better corrosion resistance.
2   When the brake assembly is being overhauled, check the condition of both the hose and the pipe for signs of leakage or scuffing, if either has made rubbing contact with the machine whilst it is in motion. The union connections at either end must also be in good condition, with no stripped threads or damaged sealing washers.

10 Front brake master cylinder: examination and renovation – SR models only

1   The master cylinder is unlikely to give trouble unless the machine has been stored for a lengthy period or until a considerable mileage has been covered. The usual signs of trouble are leakage of hydraulic fluid and a gradual fall in the fluid reservoir content.
2   To gain full access to the master cylinder, commence the dismantling operation by attaching a bleed tube to the caliper unit bleed nipple. Open the bleed nipple one complete turn, then operate the front brake lever until all fluid is pumped out of the reservoir. Close the bleed nipple, detach the tube and store the fluid in a closed container for subsequent re-use.
3   Detach the hose and also the stop lamp switch. Remove the handlebar lever pivot bolt and the lever itself.
4   Access is now available to the piston and the cylinder and it is possible to remove the piston assembly, together with all the relevant seals. Take note of the way in which the seals are arranged because they must be replaced in the same order. Failure to observe this necessity will result in brake failure.
5   Clean the master cylinder and piston with either hydraulic fluid or alcohol. On no account use either abrasive or other solvents such as petrol. If any signs of wear or damage are evident, renewal is necessary. It is not practicable to reclaim either the piston or the cylinder bore.
6   Soak the new seals in hydraulic fluid for about 15 minutes prior to fitting, then reassemble the parts **in exactly the same order,** using the reversal of the dismantling procedure. Lubricate with hydraulic fluid and make sure the feather edges of the various seals are not damaged.
7   Refit the assembled master cylinder unit to the handlebar, and reconnect the handlebar lever, hose, stop lamp etc. Refill the reservoir with hydraulic fluid and bleed the entire system by following the procedure detailed in Section 11 of this Chapter.
8   Check that the brake is working correctly before taking the machine on the road and pump the brake a few times to restore pressure and align the pads.

11 Bleeding the hydraulic system

1   As mentioned earlier, brake action is impaired or even rendered inoperative if air is introduced into the hydraulic system. This can occur if the seals leak, the reservoir is allowed to run dry or if the system is drained prior to the dismantling of any component part of the system. Even when the system is refilled with hydraulic fluid, air pockets will remain and because air will compress, the hydraulic action is lost.
2   Check the fluid content of the reservoir and fill almost to the top. Remember that hydraulic brake fluid is an excellent paint stripper, so beware of spillage, especially near the petrol tank.
3   Place a clean glass jar below the brake caliper unit and attach a clear plastic tube from the caliper bleed screw to the container. Place some clean hydraulic fluid in the container so that the pipe is always immersed below the surface of the fluid.
4   Unscrew the bleed screw one complete turn and apply the operating lever slowly. As the fluid is ejected from the bleed screw the level in the reservoir will fall. Take care that the level does not drop too low whilst the operation continues, otherwise air will re-enter the system, necessitating a fresh start.
5   Continue the pumping action with the lever until no further air bubbles emerge from the end of the plastic pipe. Hold the brake lever against the handlebars and tighten the caliper bleed screw. Remove the plastic tube AFTER the bleed screw is closed.
6   Check the brake action for sponginess, which usually denotes there is still air in the system. If the action is spongy, continue the bleeding operation in the same manner, until all traces of air are removed.
7   Bring the reservoir up to the correct level of fluid and replace the diaphragm, sealing gasket and cap. Check the entire system for leaks. Recheck the brake action.
8   Note that fluid from the container placed below the brake caliper unit whilst the system is bled should not be reused, as it will have become aerated and may have absorbed moisture.

12 Front wheel bearings: examination and replacement

All models
1   Place the machine on the centre stand or on blocks and remove the front wheel as described in Section 4. On disc brake machines remove the speedometer gearbox, drive gear and oil seal on the right-hand side of the hub and remove the spacer, dust cover and oil seal from the left-hand side. To give access to the bearings on drum brake machines, first remove the brake plate on the left-hand side of the hub and the dust excluder and oil seal from the right-hand side of the hub.
2   The wheel bearings can now be tapped out from each side with the use of a suitable long drift, passed through from the opposite side of the hub. Remove the right-hand bearing first; the bearing spacer must be knocked to one side so that the drift can be placed against the bearing race. Careful and even tapping will prevent the bearing 'tying' and damage to the races.
3   Remove all the old grease from the hub and bearings, giving the latter a final wash in petrol. When the bearings are dirty, lubricate them sparingly with a very light oil. Check the bearings for play and roughness when they are spun by hand. All used bearings will emit a small amount of noise when spun but they should not chatter or sound rough. If there is any doubt about the conditions of the bearings they should be renewed.
4   Before replacing the bearings pack them with high melting point grease. Do not overfill the hub centre with grease as it will expand when hot and may find its way past the oil seals. The hub space should be about ⅔ full of grease. Drift the bearings in, using a soft drift on the outside ring of the bearing. Do not drift the centre ring of the bearing or damage will be incurred. Replace the oil seals carefully, drifting them into place with a thick walled tube of approximately the same dimension as the oil seal. A large socket spanner is ideal.

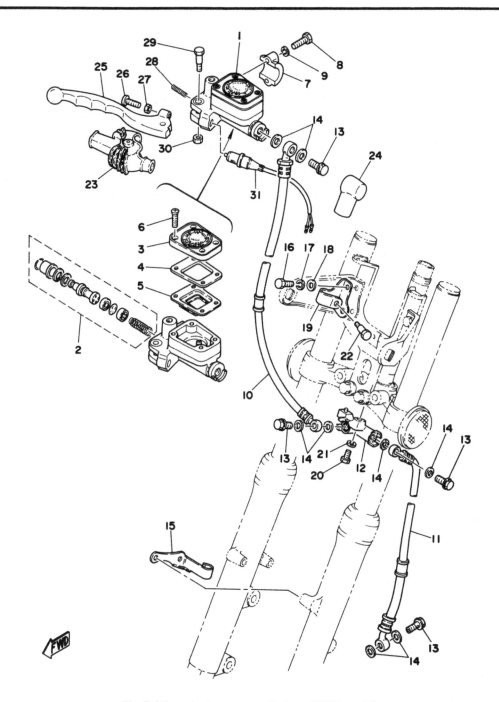

Fig. 5.4 Front brake master cylinder – SR500 models

1 Master cylinder assembly	12 Brake hose junction	22 Bolt
2 Piston kit	13 Bolt – 4 off	23 Brake lever cover
3 Reservoir cap	14 Washer – 8 off	24 Boot
4 Gasket	15 Hose clamp	25 Lever
5 Diaphragm	16 Bolt – 2 off	26 Adjuster screw
6 Screw – 4 off	17 Spring washer – 2 off	27 Nut
7 Clamp	18 Washer – 2 off	28 Spring
8 Bolt – 2 off	19 Clamp	29 Bolt
9 Spring washer – 2 off	20 Bolt	30 Nut
10 Upper hose	21 Spring washer	31 Front stop lamp switch assembly
11 Lower hose		

12.1a Remove dust covers, spacers and ...

12.1b ... any oil seals to allow ...

12.2 ... access to the wheel bearings; note seal side is outwards

12.4 Do not omit bearing spacer on reassembly

13 Rear wheel: examination, removal and renovation

1   Place the machine on the centre stand so that the rear
wheel is raised clear of the ground. Check for rim alignment,
damage to the rim and loose or broken spokes by following the
procedure relating to the front wheel, as described in Section 2
or 3 of this Chapter, depending on the type of wheel used.
2   Commence rear wheel removal by disconnecting the final
drive chain at the spring link. After separating the chain, refit the
spring link to one free end of the chain, to avoid loss. On drum
rear brake models unscrew the brake adjuster nut from the
brake rod and depress the brake pedal so that the rod leaves the
trunnion in the operating arm. Store the nut and trunnion on the
brake rod, to prevent loss. Disconnect also the brake torque rod
from the back plate. The rod is secured by a nut, bolt and split
pin.
3   Remove the split pin (where fitted) from the wheel spindle
end and undo the nut. Support the weight of the wheel and
withdraw the spindle. The chain adjuster on each fork end will
probably fall free at this stage, together with the wheel right-
hand spacer. This latter component must be removed in any

case before the wheel can be removed.
4   Replace the rear wheel by reversing the removal procedure.
On disc brake models ensure that the disc enters the brake pads
squarely so that they are not displaced or chipped. Reconnect
the final drive chain before tightening the wheel spindle nut. The
spring link is most easily fitted with the chain ends meshed on
adjacent teeth on the rear wheel sprocket. Ensure that the
closed end of the spring link clip faces the direction of normal
chain travel.
5   On drum brake models it is vital that the torque rod is
securely connected at both ends and that the nuts are
additionally secured by split pins. If the rod becomes free from
the brake plate in service, the wheel will lock on the first
application of the brake.

14 Rear drum brake: examination and renovation

1   After removal of the rear wheel as described in the preceding Section the rear drum brake should be examined and
renovated by following the procedure given for the front wheel
drum brake in Section 5.

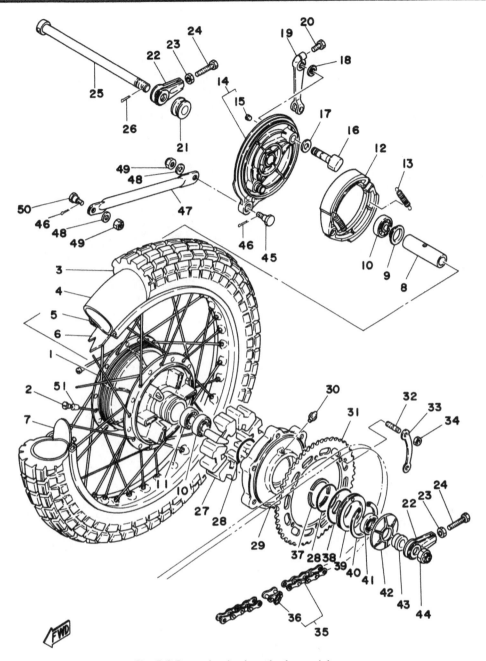

Fig. 5.5 Rear wheel – drum brake models

1	Rear wheel hub	18	Cam seal	35	Chain
2	Spoke set	19	Brake operating arm	36	Chain link
3	Tyre	20	Pinch bolt	37	Half collar – 2 off
4	Inner tube	21	Collar	38	'O' ring
5	Rear wheel rim	22	Chain adjuster – 2 off	39	Dust seal
6	Rim band	23	Nut – 2 off	40	Circlip
7	Bead spacer – 2 off	24	Bolt – 2 off	41	Oil seal
8	Bearing spacer	25	Wheel spindle	42	Dust cover
9	Dust seal	26	Split pin	43	Collar
10	Bearing – 2 off	27	Cush drive rubber – 6 off	44	Castellated nut
11	Bearing	28	'O' ring – 2 off	45	Bolt
12	Brake shoe – 2 off	29	Cush drive plate	46	Split pin – 2 off
13	Brake shoe spring – 2 off	30	Grease nipple	47	Tension arm
14	Brake plate	31	Rear wheel sprocket	48	Plain washer – 2 off
15	Grommet	32	Stud – 6 off	49	Domed nut – 2 off
16	Brake operating cam	33	Lock washer – 3 off	50	Bolt
17	Shim	34	Nut – 6 off	51	Wheel balancer

13.4a Ensure wheel spacers and chain adjuster are correctly fitted

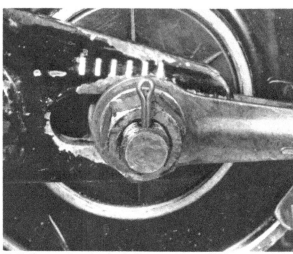

13.4b Secure the wheel spindle nut with a split pin

13.4c Ensure that chain link closed end faces direction of travel

13.5a Fit brake torque arm nut and bolt

13.5b A new split pin **must** be used to secure the nut

15 Rear disc brake: pad removal, and caliper examination

1   The rear disc brake fitted to SR500E models is almost identical to the disc front brake fitted to all the SR500 models. The procedure for inspection and removal of the pads and the method of caliper overhaul is in most respects identical to that adopted for the front brake. Refer to Sections 6 and 7 of this Chapter.

2   When renewing the brake pads, the main body of the caliper may be lifted off the support bracket without the need to disturb the bracket in any way. This may be accomplished after removing the single bolt on which the caliper slides.

16 Rear brake master cylinder: removal, examination and renovation

1   The rear brake master cylinder is mounted in board of the frame right-hand triangulation, and is so placed that the fluid level can be seen readily without the need to remove the side cover. The master cylinder is operated by a foot pedal via an adjustable push rod connected to the pedal shaft by a clevis pin and a split pin.

2   Drain the master cylinder and reservoir, using a similar technique to that described for the front brake master cylinder. The master cylinder reservoir is fitted with a triangular cap, secured by three screws, or in some cases a screw cap.

3   Disconnect the hydraulic hose at the master cylinder by removing the banjo bolt. Take care not to drop any residual fluid on the paintwork. The master cylinder is retained on the frame lug by two bolts. After removal of the bolt, the cylinder unit may be lifted upwards so that the operating push rod leaves the cylinder. The master cylinder can now be lifted away from the machine.

4   Examination and dismantling of the rear brake master cylinder may be made by referring to the directions in Section 10 of this Chapter. Additionally, the reservoir should be flushed

out with clean fluid before refitting.

17 Rear wheel bearings: removal and replacement

1   All models, except the TT500 models, have two wheel bearings on the left-hand side of the hub and a single bearing on the right-hand side. TT500 models have an additional bearing on the right-hand side to compensate for the extra loading associated with continual off-road riding.

2   The procedure for removal, examination and replacement of the rear wheel bearings is similar to that given for front wheel bearings in Section 12 of this Chapter.

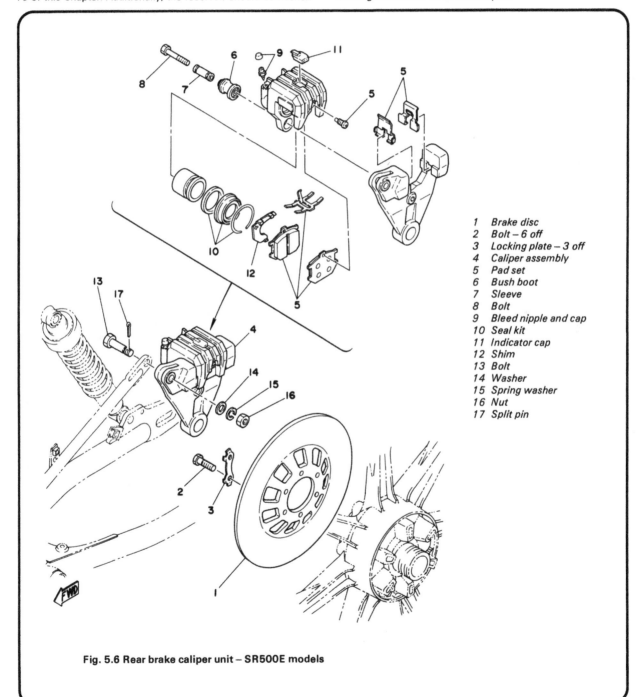

1   Brake disc
2   Bolt – 6 off
3   Locking plate – 3 off
4   Caliper assembly
5   Pad set
6   Bush boot
7   Sleeve
8   Bolt
9   Bleed nipple and cap
10 Seal kit
11 Indicator cap
12 Shim
13 Bolt
14 Washer
15 Spring washer
16 Nut
17 Split pin

Fig. 5.6 Rear brake caliper unit – SR500E models

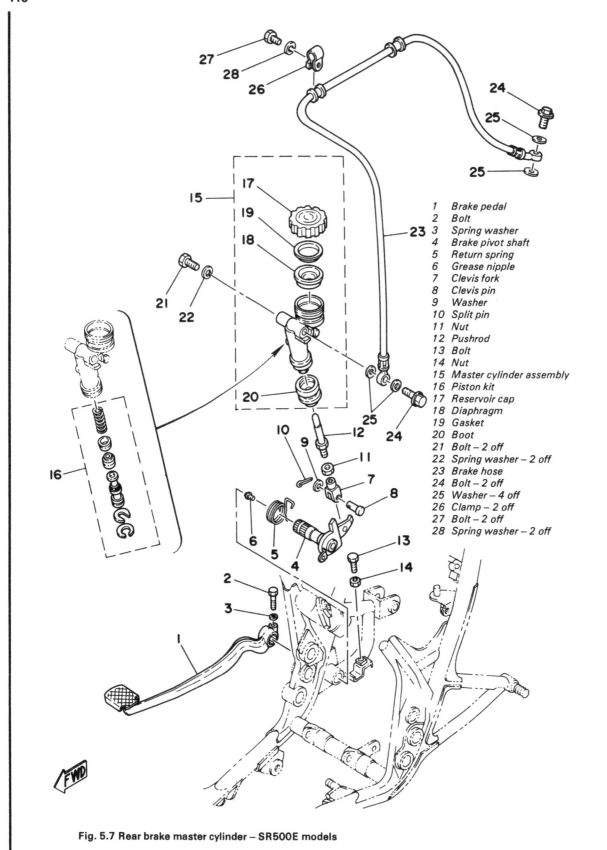

1 Brake pedal
2 Bolt
3 Spring washer
4 Brake pivot shaft
5 Return spring
6 Grease nipple
7 Clevis fork
8 Clevis pin
9 Washer
10 Split pin
11 Nut
12 Pushrod
13 Bolt
14 Nut
15 Master cylinder assembly
16 Piston kit
17 Reservoir cap
18 Diaphragm
19 Gasket
20 Boot
21 Bolt – 2 off
22 Spring washer – 2 off
23 Brake hose
24 Bolt – 2 off
25 Washer – 4 off
26 Clamp – 2 off
27 Bolt – 2 off
28 Spring washer – 2 off

FWD

Fig. 5.7 Rear brake master cylinder – SR500E models

18 Rear wheel sprocket: removal and examination

1 The rear wheel sprocket is retained on the cush drive assembly hub, or directly to the hub where no cush drive is fitted (TT models) by six studs and nuts. The nuts are secured in pairs by common tab washers.

2 Examine the sprocket for chipped, hooked or worn teeth. If renewal is necessary unbend the tab washers and remove the six retaining nuts. It is a good policy to renew the tab washers when fitting the new sprocket.

3 If a new rear sprocket is required it is almost certain that the engine drive sprocket will also require replacement. In any case, it is generally considered good practice to renew the two sprockets and the rear chain at the same time.

19 Rear wheel cush drive: examination and renovation – XT and SR models

1 The cush drive assembly consists of six rubber buffers which are housed between the rear hub and the cush drive flange casting. Six blocks cast on the inside of the drive flange transmit power from the sprocket to the rubber blocks which are held against webs cast in the wheel hub. In this way a limited amount of controlled movement is allowed between the sprocket and the rear hub, which cushions out any roughness or surging transmitted by the engine.

2 When the rear wheel is removed, it is advisable to examine the rubber buffers for signs of damage or deterioration which might render them ineffective. After extended service the rubber buffers will become permanently compacted, giving rise to excess sprocket/hub movement.

3 To gain access to the cush drive, prise off the dust cover from the left-hand side of the hub. This will give access to the large internal circlip which locates the cush drive hub. Displace the circlip and remove the hub dust seal and 'O' ring, the smaller 'O' ring and the two half collars which locate with the groove in the wheel hub and which secure the cush drive hub in position. The cush drive hub can now be lifted up to expose the rubber blocks.

4 The assembly may be reassembled by reversing the dismantling procedure.

18.2a The sprocket is mounted on six studs

18.2b Always bend up the locking plates to secure the nuts

19.3a Prise off the webbed dust cover to allow ...

19.3b ... removal of the internal circlip

19.3c Lift out the dust seal and O ring

19.3d Displace the smaller O ring and ...

19.3e ... separate the two half collars

19.3f Lift off the cush drive hub to gain access to ...

19.3g ... the cush drive rubbers

19.4a Do not omit the inner O ring on reassembly

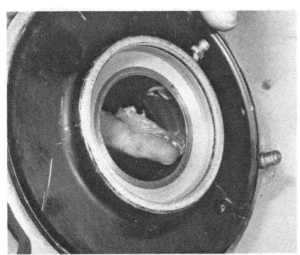

19.4b Apply grease to the hub bore before refitting

20 Rear brake pedal height: adjustment

1 The pivot shaft upon which the rear brake pedal is mounted is splined to allow adjustment of the pedal height to suit individual requirements.

2 To adjust the height, loosen and remove the pinch bolt which passes into the rear of the pedal box. Draw the pedal off the splines and refit it at the required angle. Ideally the pedal should be fitted, so that it is positioned just below the rider's right foot, when the rider is seated normally. In this way the foot does not have to be lifted before the brake can be applied.

3 The upper limit of travel of the brake pedal may be adjusted by means of the bolt and locknut fitted to a lug on the frame to the rear of the pivot shaft crank. On disc brake models care should be exercised when lowering the pedal by this method as movement imparted to the master cylinder piston may actuate the brake to a small degree. Adjustment of the pedal may necessitate readjustment of the rear stop lamp switch.

21 Final drive chain: examination and lubrication

1 The final drive chain is exposed for most of its travel and has only a lightweight chainguard to protect the upper run. No provision is made for lubricating the chain.

2 The chain tension will require adjustment at regular intervals, especially if the machine is used for competition or off-road riding. This is accomplished by slackening the rear wheel nut after first removing the split pin through the end of the spindle, and then moving the rear wheel backwards by unscrewing the two chain adjusting bolts that bear on the wheel spindle. The locknut of each adjuster should be slackened first, and retightened after adjustment has been effected. Turn each adjuster an equal amount, using the scribe marks on the fork ends to provide a visual check. Note that it will be necessary to slacken the rear brake torque arm connection during this operation and retighten it afterwards.

3 Chain tension is correct if the chain can be moved upwards and downwards 15-20 mm (0.6-0.8 in) on SR models, and 30-40 mm (1.2-1.6 in) on XT models, in the middle of the lower run. Do not run the chain overtight to compensate for uneven wear. A tight chain will place excessive stresses on the gearbox and rear wheel bearings, leading to their early failure. It will also absorb a surprising amount of power.

4 After a period of running, the chain will require lubrication. Lack of oil will accelerate the rate of wear of both chain and sprockets and will lead to harsh transmission. The application of engine oil will act as a temporary expedient, but it is preferable to remove the chain and immerse it in a molten lubricant such as Linklyfe or Chainguard after it has been cleaned in a paraffin bath. These latter lubricants achieve better penetration of the chain links and rollers and are less likely to be thrown off when the chain is in motion. Some chains are now marketed that are specially pre-lubricated and will run for much longer periods without need for attention. Refer to the chain manufacturer's recommendations in this respect.

5 Remember that if the machine is used for competition work or in particularly dusty conditions, the chain will require much more frequent attention if it is not to wear rapidly.

6 To check whether the chain is due for replacement, lay it lengthwise in a straight line and compress it endwise until all play is taken up. Anchor one end, then pull in the opposite direction to take up the play which develops. If the chain extends by more than $\frac{1}{4}$ inch per foot, it should be replaced in conjunction with the sprockets. Note that this check should ALWAYS be made after the chain has been washed out, but before any lubricant is applied, otherwise the lubricant may take up some of the play.

7 If desired, wheel alignment can be checked by running a plank of wood parallel to the machine so that it is equidistant from either side of the front wheel tyre when tested on both sides of the rear wheel. It will not touch the front tyre because

21.2 Marks on fork ends aid wheel alignment

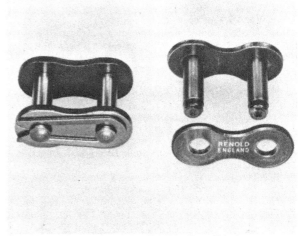

21.9 A replacement chain is offered by Renold Ltd

this tyre has a smaller cross section. See the accompanying diagram.

8   When replacing the chain, make sure that the spring link is seated correctly, with the closed end facing the direction of travel.

9   Replacement chains are now available in standard metric sizes from Renold Limited, the British chain manufacturer. When ordering a new chain, always quote the size, the number of chain links and the type of machine to which the chain is to be fitted.

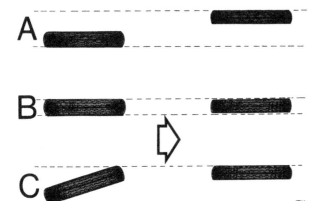

Fig. 5.8 Method of checking wheel alignment

A & C – Incorrect   B – Correct

22 Tyres: removal and replacement

1   At some time or other the need will arise to remove and replace the tyres, either as the result of a puncture or because a replacement is required to offset wear. To the inexperienced, tyre changing represents a formidable task yet if a few simple rules are observed and the technique learned, the whole operation is surprisingly simple.

2   To remove the tyre from either wheel, first detach the wheel from the machine by following the procedure in Section 4 or 13 of this Chapter depending on which wheel is involved. Deflate the tyre by removing the valve insert and when it is fully deflated, push the bead of the tyre away from the wheel rim on both sides so that the bead enters the centre well of the rim. Remove the locking cap and push the tyre valve into the tyre itself. Where security bolts are used these too should be loosened and pushed back into the tyre.

3   Insert a tyre lever close the the valve and lever the edge of the tyre over the outside of the wheel rim. Very little force should be necessary; if resistance is encountered it is probably due to the fact that the tyre beads have not entered the well of the wheel rim all the way round the tyre.

4   Once the tyre has been edged over the wheel rim, it is easy to work around the wheel rim so that the tyre is completely free on one side. At this stage, the inner tube can be removed.

5   Working from the other side of the wheel, ease the other edge of the tyre over the outside of the wheel rim that is furthest away. Continue to work around the rim until the tyre is free completely from the rim.

6   If a puncture has necessitated the removal of the tyre, reinflate the inner tube and immerse it in a bowl of water to trace the source of the leak. Mark its position and deflate the tube. Dry the tube and clean the area around the puncture with a petrol soaked rag. When the surface has dried, apply the rubber solution and allow this to dry before removing the backing from the patch and applying the patch to the surface.

7   It is best to use a patch of the self-vulcanising type which will form a very permanent repair. Note that it may be necessary

to remove a protective covering from the top surface of the patch, after it has sealed in position. Inner tubes made from synthetic rubber may require a special type of patch and adhesive if a satisfactory bond is to be achieved.

8   Before replacing the tyre, check the inside to make sure the agent which caused the puncture is not trapped. Check also the outside of the tyre, particularly the tread area, to make sure nothing is trapped that may cause a further puncture.

9   If the inner tube has been patched on a number of past occasions, or if there is a tear or large hole, it is preferable to discard it and fit a replacement. Sudden deflation may cause an accident, particularly if it occurs with the front wheel.

10 To replace the tyre, inflate the inner tube sufficiently for it to assume a circular shape but only just. Then push it into the tyre so that it is enclosed completely. Lay the tyre on the wheel at an angle and insert the valve through the rim tape and the hole in the wheel rim. Attach the locking cap on the first few threads so that it holds the valve captive in its correct location.

11 Starting at the point furthest from the valve, push the tyre bead over the edge of the wheel rim until it is located in the central well. Continue to work around the tyre in this fashion until the whole of one side of the tyre is on the rim. It may be necessary to use a tyre lever during the final stages.

12 Make sure there is no pull on the tyre valve and again commencing with the area furthest from the valve, ease the other bead of the tyre over the edge of the rim. Finish with the area close to the valve, pushing the valve up into the tyre until the locking cap touches the rim. This will ensure the inner tube is not trapped when the last section of the bead is edged over the rim with a tyre lever.

13 Check that the inner tube is not trapped at any point. Reinflate the inner tube, and check that the tyre is seating correctly around the wheel rim. There should be a thin rib moulded around the wall of the tyre on both sides which should be equidistant from the wheel rim at all points. If the tyre is unevenly located on the rim, try bouncing the wheel when the tyre is at the recommended pressure. It is probable that one of the beads has not pulled clear of the centre well.

14 Always run the tyres at the recommended pressures and never under or over-inflate. The correct pressures for solo use are given in the Specifications Section of this Chapter. If a pillion passenger is carried, increase the rear tyre pressure by approximately 4 psi and the front tyre pressure by 2 psi.

15 Tyre replacement is aided by dusting the side walls, particularly in the vicinity of the beads, with a liberal coating of French chalk. Washing up liquid can also be used to good effect, but this has the disadvantage of causing the inner surfaces of the wheel rim to rust.

16 Never replace the inner tube and tyre without the rim tape in position. If this precaution is overlooked there is good chance of the ends of the spoke nipples chafing the inner tube and causing a crop of punctures.

17 Never fit a tyre which has a damaged tread or side walls. Apart from the legal aspects, there is a very great risk of a blowout, which can have serious consequences on any two-wheel vehicle.

18 Tyre valves rarely give trouble, but it is always advisable to check whether the valve itself is leaking before removing the tyre. Do not forget to fit the dust cap which forms an effective second seal.

   Note: tyre fitting and removal   chart on page 111.

23 Tyre valve dust caps

1   Tyre valve dust caps are often left off when a tyre has been replaced, despite the fact that they serve an important two-fold function. Firstly, they prevent dirt or other foreign matter from entering the valve and causing the valve to stick open when the tyre pump is next applied. Secondly, they form an effective second seal so that in the event of the tyre valve sticking, air will not be lost.

2   Isolated cases of sudden deflation at high speeds have been traced to the omission of the dust cap. Centrifugal force has

tended to lift the tyre valve off its seating and because the dust cap is missing, there has been no second seal. Racing inner tubes contain provision for this happening because the valve inserts are fitted with stronger springs, but standard inner tubes do not, hence the need for the dust cap.

3   Note that when a dust cap is fitted for the first time, the wheel may have to be rebalanced.

24 Security bolt: function and fitting

1   If the drive from a high-powered engine is applied suddenly to the rear wheel of a motorcycle, wheel spin will occur with an initial tendency for the wheel rim to creep in relation to the tyre and inner tube. Under these circumstances there is risk of the valve being torn from the inner tube, causing the tyre to deflate rapidly, unless movement between the rim and tyre can be restrained in some way. A security bolt fulfills this role in a simple and effective manner, by clamping the bead of the tyre to the well of the wheel rim so that any such movement is no longer possible.

2   Two security bolts are fitted to the rear wheel of the XT and TT500 models. Before attempting to remove or replace a tyre the security bolts must be slackened off completely so that the clamping action is released. The inside edge of the wheel rims is ribbed to help hold the tyre firmly in position.

25 Front wheel balancing

1   The front wheel should be statically balanced, complete with tyre. An out of balance wheel can produce dangerous wobbling at high speed.

2   Some tyres have a balance mark on the sidewall. This must be positioned adjacent to the valve. Even so, the wheel still requires balancing.

3   With the front wheel clear of the ground, spin the wheel several times. Each time, it will probably come to rest in the same position. Balance weights should be attached diametrically opposite the heavy spot, until the wheel will not come to rest in any set position, when spun.

4   Machines fitted with cast aluminium wheels require special balancing weights which are designed to clip onto the centre rim flange, much in the way that weights are affixed to car wheels. When fitting these weights, take care not to affix any weight nearer than 40 mm (1·54 in) to the radial centre line of any spoke. Refer to the accompanying diagram.

5   It is possible to have a wheel dynamically balanced at some dealers. This requires its removal.

6   There is no need to balance the rear wheel under normal road conditions, although any tyre balance mark should be aligned with the valve.

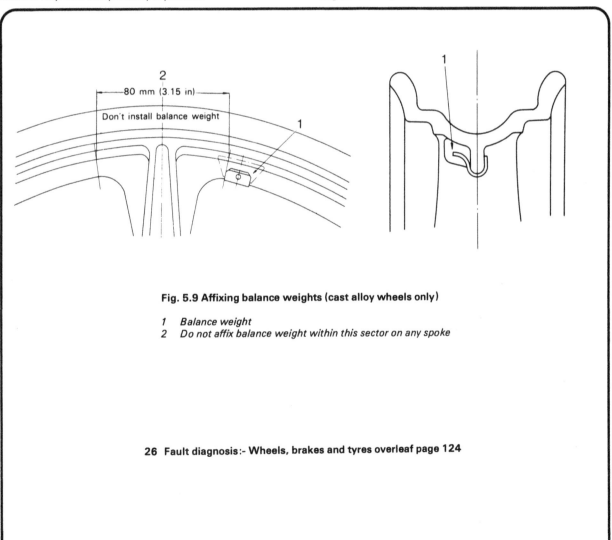

Fig. 5.9 Affixing balance weights (cast alloy wheels only)

1   Balance weight
2   Do not affix balance weight within this sector on any spoke

26 Fault diagnosis:- Wheels, brakes and tyres overleaf page 124

26 Fault diagnosis – wheels, brakes and tyres

Symptom	Cause	Remedy
Handlebars oscillate at low speeds	Buckle or flat in wheel rim, most probably front wheel	Check rim alignment by spinning wheel. Correct by retensioning spokes or having wheel rebuilt on new rim. (XT and TT models). Renew wheel (SR models).
	Tyre not straight on rim	Check tyre alignment.
Machine lacks power and accelerates poorly	Brakes binding	Hot brake drums provide best evidence. Readjust brakes (drum brakes). Caliper slide bolt binding, remove and lubricate bolt and bush.
Brakes grab when applied gently	Ends of brake shoes not chamfered Elliptical brake drum (drum brakes)	Chamfer with file. Lightly skim in lathe (specialist attention needed).
Brake pull-off sluggish	Brake cam binding in housing Weak brake shoe springs	Free and grease. Replace, if brake springs not displaced.
Harsh transmission	Worn or badly adjusted chains Hooked or badly worn sprockets	Adjust or replace as necessary. Replace as a pair, together with chain.

Chapter 6 Electrical system

Refer to Chapter 7 for information relating to the 1979 to 1983 models

Contents

Specifications

	XT500C, D and E	SR500
Battery		
Make	GS	GS
Type	6N6–3B	12N7–3B
Capacity	6V 6Ah	12V 7Ah
Earth	Negative	Negative
Flywheel generator		
Make	Nippon Denso	Nippon Denso
Type	Two-coil generator	Multi-coil alternator
Output:	6 volts	12 volts
Lighting coil resistance	0.155 ohms ± 10% at 20°C	Stator coil 0.73 ohms ± 30% at 20°C
Charging coil resistance	0.247 ohms ± 10% at 20°C	Stator coil 0.80 ohms ± 30% at 20°C
Rectifier (XT500)		
Type	Single element, half wave	
Capacity	4A	
Regulator (XT500)		
Type	Solid state	
Regulator voltage	6.9 – 7.5 volts	
Rectifier/regulator (SR500)		
Type	Integrated circuit, three phase, full-wave	
Rectifier capacity	15A	
Regulating voltage	14.5 ± 0.5 volts	

	XT500 C, D and E	SR500
Bulbs		
Headlamp	30/30W 6V (XT500E 35/35W)	50/40W 12V
Tail/stop lamp	5.3/17W 6V (XT500E 5.3/25W)	8/27W 12V
Direction indicators	17W 6V x 4	27W 12V x 4
Indicator warning	3W 6V	3.4W 12V
Instrument light	3W 6V x 2	3.4W 12V x 4
High beam warning	3W 6V	3.4W 12V
Neutral indicator	3W 6V	3.4W 12V

1 General description

The XT 500 models are fitted with a flywheel generator containing two separate power coils; one to provide ignition source power and the other to provide a lighting and charging current.The two coils are not interconnected in any way, and for the purposes of testing and fault isolating may be considered as separate component systems.

The charging coil has two output leads, both of which provide a 6 volt ac (alternating current) current. One wire provides ac current for direct operation of the headlamp and the instrument illumination lamps. This allows the machine to be used off the road without a battery. An ac regulator is included in this circuit to ensure that the voltage remains within specified limits at all times. The second circuit provides power for the tail light and ancillary equipment such as the flashing indicators and the horn. In addition, it is used to provide charging power to the battery. To enable the battery to be charged the ac current is converted to dc current by a single element half-wave silicon rectifier.

Electrical power on SR500 models is provided by a 12 volt, multi-coil alternator attached to the extreme left-hand end of the crankshaft. The alternator also incorporates the power source and timing coils for the CDI ignition system. For the purposes of fault tracing and component testing, these may be considered as a separate system. They are discussed in Chapter 3.

The ac current provided by the alternator is converted to dc current to enable the lights to be fed and the battery charged. The rectifier used for this purpose is an integrated circuit, three phase full-wave unit. The output voltage from the alternator is maintained within the range 14–15 volts by an integrated circuit regulator.

2 Checking the electrical system: general

Many of the test procedures applicable to motorcycle electrical systems require the use of test equipment of the multimeter type. Although the tests themselves are quite straightforward, there is a real danger, particularly on alternator systems, of damaging certain components if wrong connections are made. It is recommended, therefore, that no attempt be made to investigate faults in the charging system, unless the owner is reasonably experienced in the field. A qualified Yamaha Service Agent will have in his possession the necessary diagnostic equipment to effect an economical repair.

3 Flywheel generator: checking the output – XT models

1   As discussed in the general description, two types of power output are provided by the flywheel generator. Although issuing from the same generator coil, the two circuits are separate, and these should be considered separately when tracing faults. If the headlamp and instrument warning lights do not work or if the performance of the light is poor the ac circuit should be checked. If poor battery charging occurs or if the ancillary equipment functions incorrectly, it should be the dc circuit that is examined. Apart from failure of the charging coil itself, direct failure of the components involved may be due to chafed or broken wires, poor earthing or bad connections.

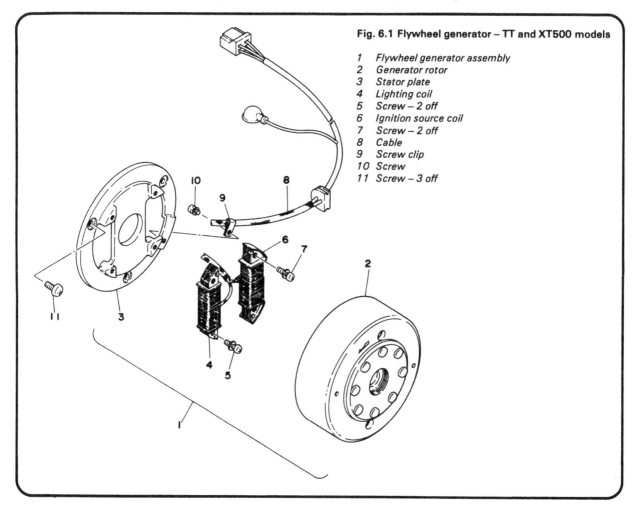

Fig. 6.1 Flywheel generator – TT and XT500 models

1   Flywheel generator assembly
2   Generator rotor
3   Stator plate
4   Lighting coil
5   Screw – 2 off
6   Ignition source coil
7   Screw – 2 off
8   Cable
9   Screw clip
10 Screw
11 Screw – 3 off

dc output and resistance test.

2   Remove the left-hand frame side cover so that access may be gained to the battery. Connect an ammeter with a 0-10A range between the positive battery terminal and the red lead, and connect a dc voltmeter of 0-20 volts range across the battery terminals. Start the engine and check both the ammeter and voltmeter readings at various engine speeds within the safe operating range.After taking the first set of readings switch the light on and repeat the tests. The results should be as shown on the accompanying chart at any given speed.

3   If the readings do not conform closely with those given on the chart or if the readings are inconsistent the resistance of the coil should be checked. Disconnect the leads from the flywheel generator at the block connector. Connect the positive lead of an ohmmeter to the lead shown in the table below for the model in question, and the negative lead to a good earth point on the engine. The correct resistance reading is as follows:

Model	Wire colour	Resistance
XT500C and D	White	0.24 ohms ± 10% at 20°C (68°F)
XT500E	White	0.21 ohms ± 10% at 20°C (68°F)

2.4 Silicon rectifier is located to the rear of the battery carrier

If the resistance is found to be incorrect, a faulty coil is indicated, and the coil must be renewed.

4   If the dc circuit output was found to be unsatisfactory, but the coil resistance is correct, the silicon rectifer should be tested. Unfortunately specific test figures for this unit are not available; it is suggested that the machine is returned to a Yamaha Service Agent, who will have the correct test equipment to determine the condition of the rectifier.

ac lighting circuit test

5   Connect an ac voltmeter of 0-20 volts range into the circuit. The positive lead should be connected to the yellow wire at the generator lead block connector. The connector should not be separated for this test. The voltmeter negative lead should be connected to a good earth point on the engine.

6   Start the engine and turn on the lights.The correct voltage is as follows:

At 2500 rpm	6.5 volts or more
At 8000 rpm	7.6 volts or less

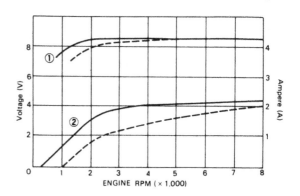

Fig. 6.2 Flywheel generator performance graph

1	Battery voltage	Day _____
2	Charging ampere	Night – – –

These voltage readings can be taken at any point in the circuit and should be the same. If the output readings are found to be incorrect, the lighting coil resistance should be checked.

7   Disconnect the wires from the generator at the block connector and connect the positive lead from an ohmmeter to the yellow lead. Connect the negative ohmmeter lead to any earth point on the engine.The lighting coil resistance should be as follows:

Model	Wire colour	Resistance
XT500C and D	Yellow	0.155 ohms ± 10% at 20°C (68°F)
XT500E	Yellow	0.170 ohms ± 10% at 20°C (68°F)

If the resistance is found to be outside the range given, the coil should be renewed. If the coil resistance is correct, but the ac circuit voltage is found to be excessive, the regulator should be checked. Unfortunately specific test equipment is required to carry out this test; for this reason it is suggested that the machine be returned to a Yamaha Service Agent who will have the correct equipment.

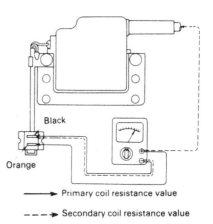

Black

Orange

———▶ Primary coil resistance value

– – –▶ Secondary coil resistance value

Fig. 6.3 Testing the ignition coil

4   Alternator: checking the output – SR500 models

1   If the performance of the charging output of the alternator is suspect, it should be checked using two successive tests, as follows: the first test should be made with a fully charged battery.

2   Connect a dc voltmeter of 0-20 volt range across the battery terminals. **Do not** disconnect the battery leads during this test, because this will damage the electrical components. Start the engine and allow it to run at 2000 rpm. At this speed and over the generated voltage should be 14-15 volts. If the output will not reach this figure, the resistance of the generating coils should be checked.

3   Disconnect the four-wire lead from the alternator at the block connector located close to the finned regulator/rectifier unit. Check the resistance across the various pairs of wires, referring to the accompanying table and illustration for specified resistances and wiring identification.

Alternator stator coil resistance

> *U-V (Yellow-White)*   *0.73 ohms ± 30% at 20°C (68°F)*
> *U-W (Yellow-White)*   *0.73 ohms ± 30% at 20°C (68°F)*
> *V-W (White-White)*   *0.80 ohms ± 30% at 20°C (68°F)*

If any reading is found to be outside the specified range, a faulty coil is indicated. Repair is impracticable and therefore the complete stator must be renewed. Before consigning the stator to the scrap bin, check that the fault is not caused by a broken or chafed wire or loose connection.

4   If, when the voltage test was taken, the output was too high, the rectifier should be checked as the next test in eliminating the faulty component. If the voltage was too low but the alternator is found to be in good condition, the rectifier should be checked.

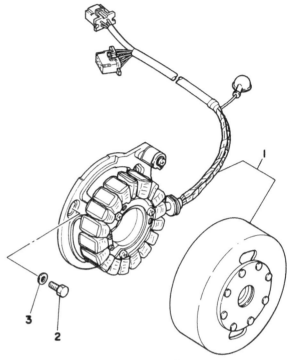

Fig. 6.4 Alternator/CDI unit

1   Generator assembly
2   Bolt – 3 off
3   Washer – 3 off

5   Rectifier/regulator unit: testing – SR500 models

1   The rectifier/regulator is a combined solid-state unit mounted behind the left-hand frame side cover forward of the battery. Although integrated as a single unit, the two functions of this component may be tested separately. If faulty performance of one function is found, the complete assembly must be renewed.

2   The rectifier can be checked whilst still in position using an ohmmeter and referring to the accompanying diagram and following table. Disconnect the rectifier block connector and check the continuity of the individual diodes. Connect the positive lead of the ohmmeter, to the first wire colour in the table and the negative lead to the second wire colour. Continue the check with each pair of wires. In all cases there should be continuity. Repeat the process with the ohmmeter leads transposed. In all cases there should be no continuity.

> *D1*   *Red (+)*   *U (white)*
> *D2*   *Red (+)*   *V (white)*
> *D3*   *Red (+)*   *W (white)*
> *D4*   *U (white)*   *Black (-)*
> *D5*   *V (white)*   *Black (-)*
> *D6*   *W (white)*   *Black (-)*

If a discrepancy is evident, the complete rectifier unit must be renewed.

3   If the regulator is functioning correctly and the alternator tests are satisfactory, but the output is found to be incorrect, by a process of elimination the regulator should be suspected as the faulty component in the system.

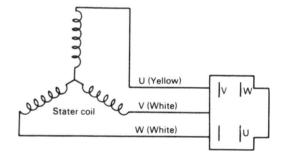

Fig. 6.5 Alternator lead identification

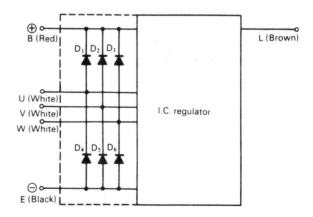

Fig. 6.6 Silicon rectifier testing

6 Battery: examination and maintenance

1   A 6 volt, 6Ah battery is fitted to all XT models, located in a carrier behind the left-hand frame cover, where it is retained by a rubber strap. On SR500 models the battery is held in place by a hinged metal strap secured by a nut and bolt; on these machines the bettery is of 12 volt, 7Ah capacity.

2   The transparent plastic case of the battery permits the upper and lower levels of the electrolyte to be observed by merely raising the battery from its housing under the dualseat. Maintenance is normally limited to keeping the electrolyte level between the prescribed upper and lower limits and making sure the vent tube is not blocked. The lead plates and their separators are visible through the transparent case, a further guide to the general condition of the battery.

3   Unless acid is spilt, as may occur if the machine falls over, the electrolyte should always be topped up with distilled water to restore the correct level. If acid is spilt onto any part of the machine, it should be neutralised with an alkali such as washing soda or baking powder and washed away with plenty of water, otherwise serious corrosion may occur. Top up with sulphuric acid of the correct specific gravity (1.260 to 1.280) only when spillage has occured. Check that the vent pipe is well clear of the frame or any of the other cycle parts.

4   If the terminals are corroded, scrape away the deposits with a sharp knife and remove the few remaining traces by wiping with a rag soaked in a strong solution of bicarbonate of soda. Do not allow any of the solution to enter the battery, since it reacts violently with the acid and may cause permanent damage to the plates. Dry off the terminals, then give them a light coating with a petroleum jelly such as Vaseline (not grease) to obviate the risk of corrosion recurring.

5   It is seldom practicable to repair a cracked battery case because the acid present in the joint will prevent the formation of an effective seal. It is always best to replace a cracked battery, especially in view of the corrosion which will be caused if the acid continues to leak.

6   If the machine is not used for a period, it is advisable to remove the battery and give it a 'refresher' charge every six weeks or so from a battery charger. If the battery is permitted to discharge completely, the plate will sulphate and render the battery useless.

7 Battery: charging procedure

1   The normal charging rate for any lead-acid battery of the type fitted to most motorcycles is 1/10 the battery capacity. The 6Ah battery will therefore have a charging rate of 0.6 amp, and the 7Ah battery that of 0.7 amp. A more rapid charge can be given in an emergency but this should be avoided if at all possible because it will shorten the useful working life of the battery. It is not advisable to exceed a charge rate of 1 ampere.

8 Fuse: location and replacement

1   A fuse within a moulded plastic case is incorporated in the electrical system to give protection from a sudden overload, such as may occur during a short circuit. It is found in close proximity to the battery, retained by a moulded rubber carrier. The fuse is rated at 10 amps (6 volt models) or 20 amps (12 volt models). A spare fuse of the same rating is carried in the rubber carrier.

2   If a fuse blows, it should not be replaced until a check has shown whether a short circuit has occurred. This will involve checking the electrical circuit to identify and correct the fault. If this precaution is not observed, the replacement fuse, which may be the only spare, may blow immediately on connection.

3   When a fuse blows whilst the machine is running and no spare is available a 'get you home' remedy is to remove the blown fuse and wrap it in silver paper before replacing it in the fuse holder. The silver paper will restore electrical continuity by bridging the broken wire within the fuse. This expedient should never be used if there is evidence of a short circuit or other major electrical fault, otherwise more serious damage will be caused. Replace the 'doctored' fuses at the earliest possible opportunity to restore full circuit protection. Always carry spare fuses.

9 Headlamp: replacing bulbs and adjusting beam height

1   In order to gain access to the headlamp bulbs it is necessary first to remove the rim, complete with the reflector and headlamp glass. The rim is retained by one of two screws passing through the headlamp shell into lugs projecting from the rim. The screw(s) are in the 8 o'clock and 4 o'clock positions viewed from the front.

2   UK models are fitted with a headlamp bulb and a pilot bulb. The headlamp bulb holder is secured to the rim by a rubber boot. After prising back the boot pull out the bulb and bulb holder. To release the bulb it should be pushed inwards slightly and twisted a number of degrees in an anti-clockwise direction. The pilot bulb holder is a bayonet fit in the reflector; the bulb too has a bayonet fitting.

3   US models have a sealed beam headlamp unit, with no

6.2 Electrolyte level is easily seen through transparent casing

8.1 Fuse is located in plastic holder together with a spare

provision for a pilot bulb. If one filament blows the complete unit must be renewed. Pull off the socket from the rear of the reflector unit to free the rim/reflector unit from the machine. To release the unit, remove the horizontal beam adjustment screw and the upper and lower pivot screws. Separate the two light unit mounting rings by removing the two screws. The light unit can be lifted out after separation. Make a note of the setting of the adjustment screw otherwise it will be necessary to re-adjust the setting after installing the new light unit.

4   Beam height on all models is effected by tilting the headlamp about the two mounting bolts. The bolts should be loosened slightly before adjustments are carried out. Horizontal beam adjustment is provided for on US models by the adjuster screw which passes through the nearside of the rim.

5   UK lighting regulations stipulate that the lighting system must be arranged so that the light will not dazzle a person standing in the same horizontal plane as the vehicle at a distance greater than 25 feet from the lamp, whose eye level is not less than 3 feet 6 inches above that plane. It is easy to approximate this setting by placing the machine 25 feet away from a wall, on a level road, and setting the beam height so that it is concentrated at the same height as the distance from the centre of the headlamp to the ground. The rider must be seated normally during this operation and also the pillion passenger, if one is carried regularly.

9.1 Remove the screw(s) to free the reflector unit

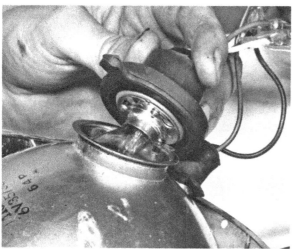

9.2a Headlamp bulb holder is retained by a rubber boot

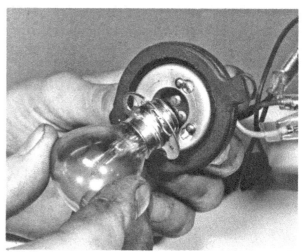

9.2b The bulb has a three-point fixing

9.2c Pilot bulb holder and bulb have bayonet fittings

10 Stop and tail lamp: replacing bulbs

1   The tail lamp has a twin filament bulb to illuminate the rear number plate and to indicate when the rear brake is applied.

2   To gain access to the stop and tail lamp bulb, unscrew the two crosshead screws which retain the plastic lens cover in position. The bulb has a bayonet fitting and also offset pins so that the stop lamp filament cannot be inadvertently connected with the tail lamp and vice versa.

3   Machines imported into countries other than the UK may have a combined stop and tail lamp assembly of greater size and fitted with a different wattage bulb to conform to the local lighting regulations. Refer to the local Yamaha Agent for details of the modified specification.

11 Flashing indicator lamps: replacing bulbs

1   Flashing indicator lamps are fitted to the front and rear of the machine on short stalks through which the electrical leads pass.

2   To gain access to the bulb in each flasher remove the lens, held by two screws. The bulb is of the bayonet fitting type. To release the bulb it should be depressed and twisted anti-clockwise a few degrees.

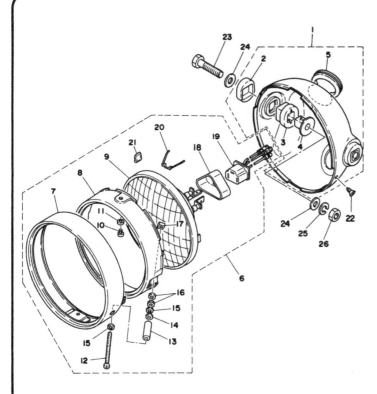

Fig. 6.7 Headlamp assembly – sealed beam type

1   Headlamp assembly
2   Grommet – 2 off
3   Grommet – 2 off
4   Collar – 2 off
5   Grommet
6   Headlamp unit
7   Headlamp rim
8   Retaining rim
9   Reflector unit
10 Screw – 2 off
11 Spring washer – 2 off
12 Adjusting screw
13 Spacer
14 Washer
15 Spring washer – 2 off
16 Nut – 2 off
17 Nut
18 Cover
19 Bulb socket
20 Spring – 3 off
21 Cover
22 Screw
23 Bolt – 2 off
24 Washer – 4 off
25 Spring washer – 2 off
26 Nut – 2 off

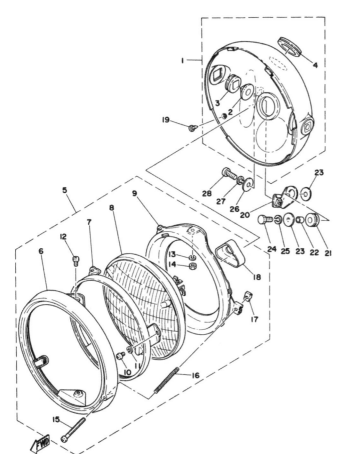

Fig. 6.8 Headlamp assembly – SR500 models

1   Headlamp shell
2   Grommet – 2 off
3   Grommet – 2 off
4   Grommet – 3 off
5   Headlamp assembly
6   Headlamp rim
7   Retaining rim
9   Mounting ring
10 Screw – 3 off
11 Spring washer – 3 off
12 Screw – 2 off
13 Spring washer – 2 off
14 Nut – 2 off
15 Adjusting screw
16 Spring
17 Nut
18 Cover
19 Rim fitting screw
20 Headlamp clamp
21 Grommet
22 Collar
23 Plain washer
24 Bolt
25 Spring washer
26 Washer
27 Spring washer
28 Screw

10.2a Tail/stop lamp lens is held by two screws

10.2b The bulb has an off-set pin bayonet fitting

11.2 Flasher lens is held by two screws

12 Flasher unit: location and replacement

1　The flasher relay unit is located either under the dualseat or behind the left-hand side cover, and is carried in a moulded rubber mounting which protects the units from vibration.

2　If the flasher unit is functioning correctly, a series of audible clicks will be heard when the indicator lamps are in action. If the unit malfunctions and all the bulbs are in working order, the usual symptom is one initial flash before the unit goes dead; it will be necessary to replace the unit complete if the fault cannot be attributed to any other cause.

3　On SR500 models, an electronic flashing indicator cancelling unit is incorporated in the indicator system. The unit automatically turns the flasher light off a certain time after the flasher switch has been operated. The time lapse is dependent on the speed of the machine. If the machine is travelling fast, the unit cancels automatically after a short time. The slower the machine is travelling, the longer time taken for cancellation. The system may be overriden manually in the normal manner.

4　Take great care when handling either unit because they are easily damaged if dropped.

13 Speedometer and tachometer heads: replacement of bulbs

1　The indicator and illuminating bulbs fitted to the speedometer and tachometer heads are rubber mounted and press into the base of each instrument. To gain access to the holders the drive cables must be disconnected from the instrument heads by unscrewing the knurled rings, and the instruments separated from their covers by removing the mounting nuts from the instrument bases. Each instrument may be attended to separately.

2　The bulbs fitted are of the standard bayonet fitting type.

14 Ignition and lighting switch

1　The ignition and lighting switch is combined in one unit, bolted to the top fork yoke. It is operated by a key which cannot be removed when the ignition is switched on. On SR500 models the switch also incorporates the steering lock.

2　If the switch proves defective, it can be removed by unscrewing the two bolts which secure it to the yoke and separating the terminal connector at the end of the short wiring harness. Remember that when a new switch if fitted, it will be necessary also to change the ignition key.

3　It is rarely possible to effect a satisfactory repair if the switch malfunctions. If trouble is experienced, the switch should be renewed.

15 Handlebar switches: general

1　Generally speaking, the switches give little trouble, but if necessary they can be dismantled by separating the halves which form a split clamp around the handlebars. Note that the machine cannot be started until the ignition cut-out on the right-hand end of the handlebars is turned to the central 'ON' position.

2　Always disconnect the battery before removing any of the switches, to prevent the possibility of a short circuit. Most troubles are caused by dirty contacts, but in the event of the breakage of some internal part, it will be necessary to renew the complete switch.

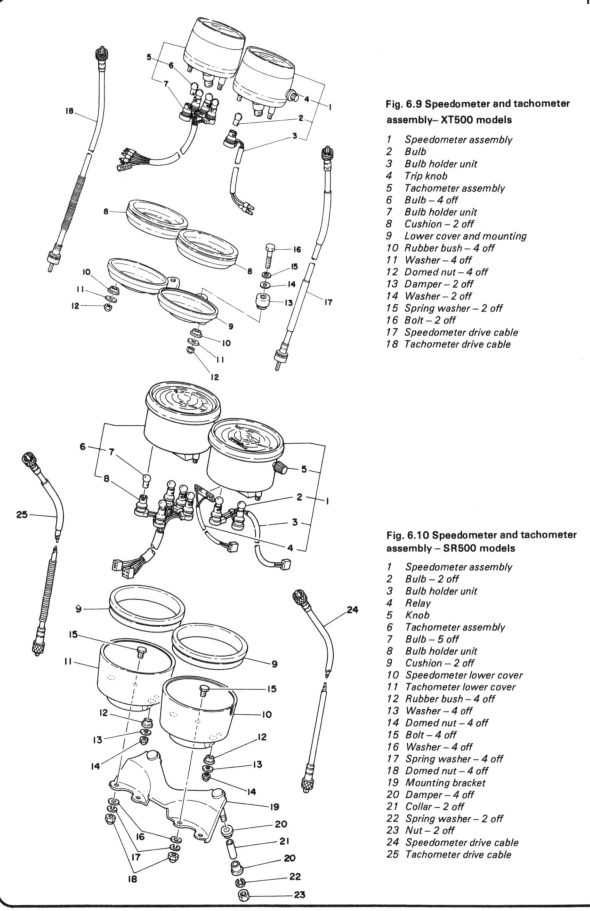

Fig. 6.9 Speedometer and tachometer assembly– XT500 models

1　Speedometer assembly
2　Bulb
3　Bulb holder unit
4　Trip knob
5　Tachometer assembly
6　Bulb – 4 off
7　Bulb holder unit
8　Cushion – 2 off
9　Lower cover and mounting
10　Rubber bush – 4 off
11　Washer – 4 off
12　Domed nut – 4 off
13　Damper – 2 off
14　Washer – 2 off
15　Spring washer – 2 off
16　Bolt – 2 off
17　Speedometer drive cable
18　Tachometer drive cable

Fig. 6.10 Speedometer and tachometer assembly – SR500 models

1　Speedometer assembly
2　Bulb – 2 off
3　Bulb holder unit
4　Relay
5　Knob
6　Tachometer assembly
7　Bulb – 5 off
8　Bulb holder unit
9　Cushion – 2 off
10　Speedometer lower cover
11　Tachometer lower cover
12　Rubber bush – 4 off
13　Washer – 4 off
14　Domed nut – 4 off
15　Bolt – 4 off
16　Washer – 4 off
17　Spring washer – 4 off
18　Domed nut – 4 off
19　Mounting bracket
20　Damper – 4 off
21　Collar – 2 off
22　Spring washer – 2 off
23　Nut – 2 off
24　Speedometer drive cable
25　Tachometer drive cable

16 Stop lamp switches: adjustment

1   All models have a stop lamp switch fitted to operate in conjunction with the rear brake pedal. The switch is located immediately to the rear of the crankcase, on the right-hand side of the machine. It has a threaded body, permitting a range of adjustment.

2   If the stop lamp is late in operating, slacken the locknuts and turn the body of the lamp in an anti-clockwise direction so that the switch rises from the bracket to which it is attached. When the adjustment seems near correct, tighten the locknuts and test.

3   If the lamp operates too early, the locknuts should be slackened and the switch body turned clockwise so that it is lowered in relation to the mounting bracket.

4   As a guide, the light should operate after the brake pedal has been depressed by about 2 cm (¾ inch).

5   SR500 models have a front stop lamp switch incorporated in the master cylinder body mounted on the handlebars. The switch is not adjustable, and if it malfunctions it must be renewed.

17 Horn

1   The horn is mounted below the headlamp in a forwards facing direction. It is bolted to a bracket that extends from the lower yoke. After considerable use the contacts inside the horn will wear. To compensate for wear an adjusting screw is fitted at the rear of the horn. If the horn tone becomes inaudible or poor, turn the screw in slowly until the tone is correct again. Do not turn the screw in too far or thecurrent increase may burn out the horn coil.

18 Wiring: layout and examination

1   The wiring is colour-coded and will correspond with the accompanying wiring diagrams.

2   Visual inspection will show whether any breaks or frayed outer coverings are giving rise to short circuits. Another source of trouble may be the snap connectors, particularly where the connector has not been pushed home fully in the outer casing.

3   Intermittent short circuits can sometimes be traced to a chafed wire which passes through a frame member. Avoid tight bends in the wire or situations where the wire can become trapped or stretched between casings.

19 Fault diagnosis:– electrical system

Symptom	Cause	Remedy
Complete electrical failure	Blown fuse	Check wiring and electrical components for short circuit before fitting new 10 amp fuse. Check battery connections, also whether connections show signs of corrosion.
Dim lights, horn inoperative	Discharged battery	Recharge battery with battery charger. Check whether generator is giving correct output.
Constantly blowing bulbs	Vibration, poor earth connection	Check security of bulb holders. Check earth return connections.

Chapter 7 The 1979 to 1983 models

Contents

Specifications

Except where entered below, specifications for the models covered in this Chapter are the same as those given for the earlier TT, XT and SR500 E models at the beginning of each Chapter.

Dimensions and weights – mm (in)	TT500	XT500 G, H	SR500 F
Overall length	2115 (83.3)	2145 (84.4)	2090 (82.3)
Overall width	875 (34.4)	875 (34.4)	835 (32.9)
Overall height	1155 (45.5)	1160 (45.7)	1120 (44.1)
Wheelbase..	1420 (55.9)	1415 (55.7)	1400 (55.1)
Minimum ground clearance	230 (9.1)	225 (8.9)	140 (5.5)
Dry weight – kg (lbs)	123 (271)	138 (304)	163 (359)
	SR500 G, H	UK XT500	UK SR500
Overall length	2105 (82.9)	2175 (85.6)	2105 (82.9)
Overall width	835 (32.9)	875 (34.3)	930 (36.6)
Overall height	1150 (45.3)	1170 (46.1)	1140 (44.9)
Wheelbase..	1410 (55.5)	1420 (55.9)	1400 (44.9)
Minimum ground clearance	165 (6.5)	220 (8.7)	165 (6.5)
Dry weight – kg (lbs)	160 (353)	139 (306)	160 (353)

Specifications relating to Chapter 1

Valve clearance – all models

Inlet..	0.10 mm (0.004 in)
Exhaust..	0.15 mm (0.006 in)

Gearbox

Final drive ratio – TT500..	3.333:1

Main torque wrench settings	kgf m	lbf ft
Engine mounting bolts – XT and SR500:		
Front and rear bolts..	4.8	34.5
Upper bolt	2.5	18.0
Engine mounting bolts – TT500:		
Rear bolts...	6.4	44.0
Upper bolt	2.5	18.0

Specifications relating to Chapter 2

Carburettor	TT500	XT500 F	XT500 G, H
Make	Mikuni	Mikuni	Mikuni
Type	VM34SS	VM32SS	VM32SS
Main jet	210	230	250
Air jet	Not available		
Jet needle	6H2-4	6FL24-3	5FL34-Fixed
Needle jet	Q-2	Q-0	P-8
Cut-away	4.0	3.5	3.5
Pilot jet	30	25	20
Starter jet	Not available		
Air screw (turns out)	Preset	Preset	Preset
Float level	22.0 ± 1 mm	22.0 ± 1 mm	23.5 ± 1 mm
Idle speed (rpm)	Not available	1200	1200

Carburettor	SR500 F, UK SR500	SR500 G, H
Make	Mikuni	Mikuni
Type	VM34SS	VM32
Main jet	300	260
Air jet	Not available	
Jet needle	6FL25-2	5FL31- Fixed
Needle jet	P-8	P-8
Cut-away	3.5	3.5
Pilot jet	25	20
Starter jet	Not available	
Air screw (turns out)	$1\frac{7}{8}$ (UK), Preset	Preset (US)
Float level	23.5 ± 1 mm	23.5 ± 0.5 mm
Idle speed (rpm)	1100	1100

Specifications relating to Chapter 3

Flywheel generator – US XT500 G, H
Pulser coil resistance:
    High speed (white to red/black) .... 16 ohms ± 10% at 20°C
    Low speed (white/green to black) .... 91 ohms ± 10% at 20°C
Source coil resistance:
    High speed (red to black) .... 390 ohms ± 10% at 20°C
    Low speed (brown to black) .... 381 ohms ± 10% at 20°C

Ignition HT coil – US XT500 G and H
Primary winding resistance .... 0.98 ohm ± 10 %
Secondary winding resistance .... 12.0 K ohm ± 20 %

Ignition timing
Retarded:
    XT500 G and H .... 7° BTDC @ 1200 rpm
    SR500 F .... 7° BTDC @ 1100 rpm
    SR500 G and H .... 12° BTDC @ 1100 rpm
Full advance:
    XT500 G and H .... 33.5° BTDC @ 7000rpm
    SR500 models .... 33.5° BTDC @ 6000 rpm
Range:
    XT500 G, H and SR500 F .... 26.5°
    SR500 G, H .... 21.5°
Advancer system – XT500 G, H .... Electronic

Spark plug type	NGK	Champion	ND
TT500 H	B7ES	N4C	W22ES-U
XT500 G, H and SR500	BP6ES	N281YC	W20EP-U

Specifications relating to Chapter 4

Front forks:
Oil capacity (per leg):
    TT500 .... 247 cc (8.31 US fl oz)
    XT500 G and H .... 238 cc (8.00 US fl oz)

Specifications relating to Chapter 4 (continued)

Spring free length:
TT500 .... 539.5 mm (21.2 in)
XT500 G and H .... 481 mm (18.9 in)
SR500 .... 445 mm (17.5 in)

Oil type:
TT500 .... SAE 20 fork oil
XT500:
UK models .... SAE 10W/30 engine oil
US models .... SAE 10 fork oil
SR500:
UK models .... SAE 20W/40 engine oil
US models .... SAE 10 fork oil

Specifications relating to Chapter 5

Tyre size

TT500 rear tyre .... 4.60 x 18

Tyre pressures

XT500 front tyre (on road) .... 21 psi (1.5 kg/cm^2)

Brakes

Minimum lining thickness:
Brake pad .... 6 mm (0.24 in)
Brake shoe .... 2 mm (0.08 in)

Specifications relating to Chapter 6

Flywheel generator – US XT500 G, H

Lighting coil resistance (yellow to black) .... 0.24 ohms ± 10% at 20°
Charging coil resistance (white to black) .... 0.28 ohms ± 10% at 20°C

Bulbs

	UK XT500	UK SR500
Headlamp	6V 35/35W	12V 60/55W
Tail/stop lamp	6V 21/5W	12V 21/5W
Direction indicators	6V 17W x 4	12V 27W x 4
Instrument light	6V 3W x 2	12V 3.4W x 2

1 Introduction

The first six chapters of this manual relate to the pre-1979 models. This chapter describes the changes made to all models from 1979 onwards.

When working on a 1979 or later model refer first to this Chapter. If the information required is not found it can be assumed that the task can be performed using the information given in the relevant Sections of Chapters 1 to 6.

To assist owners in identifying their machines correctly, the following text describes the modifications made from 1979 onwards. As a further aid to identification, the exact model can be determined using the list of engine and frame numbers at the end of this section.

UK XT500 model

The 1980 (model code 4E5) model was largely unchanged from the earlier model apart from minor cosmetic alterations. The only significant changes were in the fitting of a slightly modified front fork and minor modifications to the engine. This model continued unchanged until it was discontinued in April 1983.

UK SR500 model

A revised SR500 model (model code 4E6) was introduced in January 1981, which remained in production until discontinued in April 1983. This model was basically unchanged from its predecessor apart from minor cosmetic changes and a few minor modifications to the engine. The only change to the chassis was that the two-piece front brake hose and junction assembly was replaced by a single hose.

US XT500 models

The 1979 XT500 F model was identical to the earlier E model, apart from a few minor modifications to the engine and the usual cosmetic changes.

For 1980 the F model was superseded by the XT500 G. The contact breaker ignition system was replaced by a CDI system (similar to that fitted to SR500 models), and a slightly modified front fork was fitted.

In 1981 the XT500 H model was introduced which, apart from the changes to paintwork and graphics, was almost identical to the G model.

US SR500 models

The 1979 SR500 F is identical to the E model, apart from a change of colour and graphics, and a minor modification to the brakes.

The main change on the 1980 SR500 G model, which replaced the F, was that the hydraulically operated rear disc brake was replaced by a drum brake. The wheels were also designed to accept tubeless tyres.

In 1981 the SR500 H model was introduced which, apart from the usual cosmetic changes, was almost identical to the G model.

US TT500 models

The only major change from the E model on the 1979 TT500 F model was the fitting of a modified front fork. Apart from the usual colour and graphics changes the 1980 TT500 G model remained unchanged from the F model. The 1981 TT500 H model was also identical to the G model, apart from the usual colour and graphics change.

Model	Dates of production	Engine/frame number
UK XT500 (4E5)	Jan '80 to Apr '83	1U6-140101 on
UK SR500 (4E6)	Jan '81 to Apr '83	2J2-0252594 on
US XT500 F	1979	1E6-220101 on
US XT500 G	1980	3H6-000101 on
US XT500 H	1981	4R9-000101 on
US SR500 F	1979	2J2-020101 on
US SR500 G	1980	3H1-000101 on
US SR500 H	1981	4R8-000101 on
US TT500 F	1979	1T1-200101 on
US TT500 G	1980	1T1-230101 on
US TT500 H	1981	2Y0-000101 on

2   Routine maintenance: service schedules

Note that revised service schedules have been introduced for all later models. Routine maintenance should now be carried out at the intervals given under the relevant sub-heading. Unless otherwise stated, the task can be performed using the information in the main Routine maintenance section at the start of this manual, referring to the revised Specifications at the start of this Chapter.

All models
Daily (pre-ride) check
   Check the engine oil level.
   Check the fuel level. Ensure you have enough fuel to complete your journey or at least to get you to the nearest filling station.
   Check the wheels and tyres, especially tyre pressures and tread wear.
   Check all controls are correctly adjusted and working smoothly.
   Check that the speedometer, horn and lights are working correctly.

Every 300 miles (500 km)
   Repeat all the tasks listed under the daily checks and:
   Clean and lubricate the drive chain.

UK models
Every 1000 miles (1500 km)
   Repeat all previous maintenance tasks and carry out the following:
   Clean the air filter.
   Check the battery electrolyte level.
   Check and adjust drive chain tension.
   Check and adjust the brakes.
   Check the wheels and tyres.
   Check the tightness of all nuts, bolts and other fasteners.

Every 2000 miles (3000 km)
   Carry out all tasks listed under previous headings and:
   Change the engine oil.
   Check and adjust valve clearances.
   Check and adjust cam chain.
   Check the spark plug condition.
   Check ignition timing.
   Check carburettor settings.
   Check and adjust the clutch.

Clean fuel filter. Refer to Section 3 of Chapter 2.
   Grease the throttle twistgrip and lubricate all control cables.
   Lubricate brake pedal and stand pivots.

Every 4000 miles (6000 km)
   Repeat all tasks given under previous checks, then carry out the following:
   Change the oil filter.
   Renew the spark plug.
   Lubricate the swinging arm pivot shaft and steering head bearings.
   Remove, clean and grease the wheel bearings and the speedometer drive gearbox.
   Change front fork oil.

US XT and SR500 models
Six monthly or every 2500 miles (4000 km)
   Carry out all tasks listed under the previous headings and carry out the following:
   Change the engine oil.
   Check the condition of the spark plug.
   Check and adjust valve clearances.
   Check and adjust the cam chain.
   Clean the air filter.
   Check the ignition timing (F models only).
   Check carburettor settings.
   Check the exhaust system for leaks, renewing gaskets if necessary.
   Check and adjust the clutch.
   Check and adjust the brakes.
   Check battery electrolyte level.
   Check wheel bearings.
   Lubricate control cables and levers.
   Lubricate stand pivots.

Annually or every 5000 miles (8000 km)
   Repeat all tasks given under previous checks, then carry out the following:
   Change the oil filter.
   Renew the spark plug.
   Check crankcase ventilation system. Examine the ventilation hoses and breather chamber for cracks and splits, renewing if necessary.
   Examine the fuel pipe for damage. Check the fuel pipe for signs of cracks or splits and renew it if necessary.
   Lubricate the swinging arm pivot shaft and brake pedal pivot point referring to Chapter 4 for further information.

Two yearly or every 10 000 miles (16 000 km)
   Carry out all tasks listed under previous headings, then carry out the following:
   Change front fork oil.
   Grease the steering head bearings.

US TT500 models
Due to these machines being used off-road, it is difficult to give specific intervals for maintenance tasks to be performed. The schedule listed below should be used as a guide but with experience it may prove necessary to increase or decrease the frequency of servicing depending on the use to which the machine is to be put.

Every 300 miles (500 km)
   Clean, lubricate and adjust the drive chain.

Every 2000 miles (3000 km)
   Change the engine oil.
   Check and adjust valve clearances.
   Check the condition of the spark plug.
   Check and adjust ignition timing.
   Clean the air filter.
   Check carburettor settings.
   Check and adjust the clutch.
   Lubricate throttle twistgrip and control cables.
   Lubricate brake pedal pivot point.

Every 4000 miles (6000 km)
   Carry out all the tasks listed under the previous heading and then:
   Change the oil filter.
   Check and adjust cam chain.
   Check cylinder compression.
   Lubricate swinging arm pivot shaft and steering head bearings.
   Change the front fork oil.
   Remove, clean and grease the wheel bearings.

3   Engine, clutch and gearbox: modifications

Engine

1   With reference to Fig.1.7, note that on all later models there is a plain washer fitted between the camshaft sprocket retaining bolt and the piston position indicator (items 7 and 8).
2   During 1983 Yamaha modified the lubrication system on UK models. A gasket was fitted between the oil pump cover and crankcase (Chapter 1, Section 36), and the rocker box oil feed pipe was re-routed. The new pipe supplies oil directly to the exhaust valve rocker arm spindle on the left-hand side of the rocker box. The original point on the right-hand side is now blocked with a plug and sealing washer. In addition, the rocker arm spindles were modified, also during 1983. The new spindles have a threaded hole one end and each is retained in the rocker box by a bolt and locking tab. Before either spindle can be removed, the tab washer must be straightened and the bolt removed. On refitting, tighten the bolt securely and bend one of the tabs against one of the flats of the bolt.

Clutch

3   On all UK models and US XT, TT and SR500 H models there is an additional thrust plate fitted behind the clutch assembly. This thrust plate is fitted between the original thrust plate (Fig.1.11, item 16) and the crankcase.

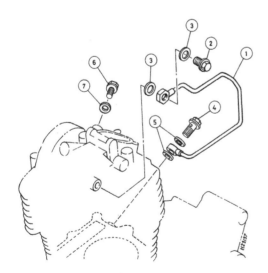

Fig. 7.1 Rocker box oil feed pipe modification – later UK models

1	Oil feed pipe	5	Sealing washer – 2 off
2	Upper union bolt	6	Blanking plug
3	Sealing washer – 2 off	7	Sealing washer
4	Lower union bolt		

Gearbox

4   The gearchange shaft and arm (items 8 and 10, Fig.1.12) are no longer two separate components. Both are fixed firmly together and cannot be separated. Therefore, the E-clip (item 11) is no longer required.

4   Ignition system: testing – US XT500 G and H models

Ignition source coil

1   The source coil is an integral part of the stator coil assembly and cannot be renewed individually. It can be tested by making resistance checks across its wiring connectors after disconnecting the main block connector from the alternator. Note that the resistance checks are made on the coil side of the block connector and should comply with those given in the Specifications. If either reading differs greatly, the coil is confirmed faulty, although check first that this is not due to a broken wire or poor connection.

Pulser coil

2   The coil can be tested without the need for its removal, by disconnecting the block connector from the alternator and checking the resistance across the coil's wires. Note that two tests should be made to check the low and high speed windings; refer to the Specifications for details of the relevant wire colours and resistance figures. Note that if the test results are greatly different from those given that coil should be renewed. Check first, however, that the fault is not due to a broken wire or poor connection. The pulser coil is mounted on the alternator stator, access being gained by removing the left-hand crankcase cover. A single screw retains the unit on the stator.

Ignition HT coil test

3   The coil can be tested as described in Chapter 3, Section 8, for the SR500 model.

CDI unit

4   The CDI unit is located near the battery, together with the regulator and rectifier units. It is a sealed unit and cannot be repaired. Furthermore, no test details are available with which it can be tested. If a fault exists in the ignition system, which cannot be traced to any other component, then the CDI unit should be considered faulty and renewed. Before renewing the unit, however, have your findings confirmed by a Yamaha dealer.

Ignition timing

5   Ignition timing should be checked as described in Chapter 3, Section 9 for the SR 500 models, noting that the F mark on the rotor should align with the index mark at an engine speed of 1200 rpm. Note that the stator plate bolt holes are slotted to permit adjustment, although it is unlikely that this should ever be necessary.

5   Front forks: modifications

All the XT and TT500 models covered in this Chapter, except the XT500 F model, are fitted with a modified front fork (see accompanying figure). This fork is similar in construction to that fitted to the earlier SR500 models, covered in Chapter 4. The only major difference concerns the fork fitted to the XT models, which has a different seal arrangement incorporating a dust seal as well as an oil seal.

Fig. 7.2 Front forks – XT models (TT models similar)

1   Cap
2   Cap bolt
3   O-ring
4   Spacer
5   Spring seat
6   Clamp
7   Gaiter
8   Clamp
9   Spring
10   Damper rod and rebound
     spring
11   Stanchion
12   Damper rod seat
13   Dust seal cover
14   Dust seal
15   Circlip
16   Oil seal clip
17   Washer
18   Oil seal
19   Left-hand lower leg
20   Sealing washer
21   Socket screw
22   Drain screw
23   Sealing washer
24   Cable guide
25   Bracket
26   Bolt
27   Washer – 5 off
28   Cable guide
29   Clamp
30   Spring washer – 4 off
31   Nut – 4 off
32   Bolt – 3 off
33   Right-hand headlamp bracket
     – XT model
34   Left-hand headlamp bracket –
     XT model

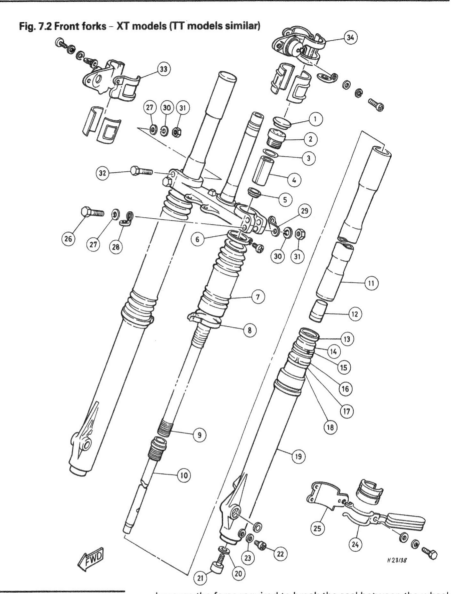

6   Wheels, brakes and tyres: modifications

1   As mentioned previously, on the UK SR500 model the original two-piece brake hose and junction assembly (Fig.5.4) fitted to the earlier models has been replaced by a single flexible hose.
2   On all SR500 models covered in this Chapter, the anti-squeal shim (item 12 in Figs. 5.3 and 5.6) is omitted from the piston side brake pad. On SR500 G and H models the rear disc brake is replaced with a drum brake. Refer to Chapter 5 for all operations involving the braking system.
3   The SR500 G and H models are fitted with wheels which can accept tubeless tyres. Apart from this, the wheels are identical to those fitted to earlier models. For operations involving tubeless tyres refer to the following Sections of this Chapter, otherwise refer to Chapter 5 for all operations involving the wheels.

7   Tubeless tyres: removal and refitting

1   It is strongly recommended that should a repair to a tubeless tyre be necessary, the wheel is removed from the machine and taken to a tyre fitting specialist or an authorized dealer. This is because the force required to break the seal between the wheel rim and tyre bead is considerable and likely to be beyond the capabilities of an individual working with normal tyre removing tools. Also, any abortive attempt to break the rim to bead seal may cause damage to the wheel rim, resulting in expensive wheel renewal. If, however, a suitable bead releasing tool is available, and experience has already been gained in its use, tyre removal and refitting can be accomplished as follows.
2   Remove the wheel from the machine. Deflate the tyre by removing the valve core and when it is fully deflated, push the bead of the tyre away from the wheel rim on both sides so that the bead enters the well of the rim. As noted, this operation will almost certainly require the use of a bead releasing tool.
3   Insert a tyre lever close to the valve and lever the edge of the tyre over the outside of the wheel rim. Very little force should be necessary; if resistance is encountered it is probably due to the fact that the tyre beads have not entered the well of the wheel rim all the way round the tyre. Should the initial problem persist, lubrication of the tyre bead and the inside edge and lip of the rim will facilitate removal. Use a recommended lubricant, a diluted solution of washing-up liquid or french chalk. Lubrication is usually recommended as an aid to tyre fitting but its use is equally desirable during removal. The risk of lever damage to wheel rims can be minimised by the use of proprietary plastic rim protectors placed over the rim flange at the point where the tyre

Fig. 7.3 Rear wheel – SR500 models

1   Wheel spindle
2   Chain adjuster – 2 off
3   Locknut – 2 off
4   Bolt – 2 off
5   Left-hand spacer
6   Oil seal
7   Circlip
8   Dust seal
9   Half collar – 2 off
10 Rear wheel sprocket
11 Nut – 6 off
12 Tab washer – 3 off
13 Stud – 6 off
14 Cush drive plate
15 Grease nipple
16 Cush drive rubbers
17 Bearing – 2 off
18 Spacer flange
19 Centre spacer
20 Bearing
21 Brake shoes
22 Return spring – 2 off
23 Brake operating cam
24 Shim
25 Brake plate
26 Grommet
27 Right-hand spacer
28 Nut
29 Split pin
30 Brake operating arm
31 Cam seal
32 Bolt
33 Nut

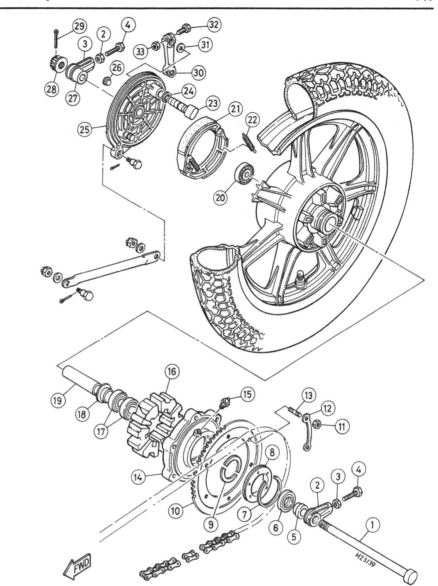

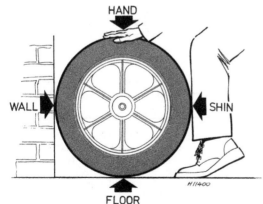

Fig. 7.4 Method of seating the beads on tubeless tyres

levers are inserted. Suitable rim protectors can be fabricated very easily from short lengths (4 – 6 inches) of thick-walled nylon petrol pipe which have been split down one side using a sharp knife. The use of rim protectors should be adopted whenever levers are used and, therefore, when the risk of damage is likely.

4   Once the tyre has been edged over the wheel rim, it is easy to work around the wheel rim so that the tyre is completely free on one side.
5   Working from the other side of the wheel, ease the other edge of the tyre over the outside of the wheel rim, which is furthest away. Continue to work around the rim until the tyre is freed completely.
6   Refer to the following sections for details of puncture repair, tyre renewal and valves.
7   Tyre refitting is virtually a reversal of the removal procedure. If the tyre has a balance mark (usually a spot of coloured paint), this must be positioned alongside the tyre valve. Similarly, any arrow indicating direction of rotation must face the right way.
8   Starting at the point furthest from the valve, push the tyre bead over the edge of the wheel rim until it is located in the well. Continue to work around the tyre in this fashion until the whole of one side of the tyre is on the rim. It may be necessary to use a tyre lever during the final stages. Here again, the use of a lubricant will aid fitting. It is strongly recommended that when fitting the tyre only a recommended lubricant is used because such lubricants also have sealing properties. Do not be over generous in the application of lubricant or tyre creep may occur.
9   Fitting the second bead is similar to fitting the first. Start by pushing the bead over the rim and into the well at a point diametrically opposite the tyre valve. Continue working around

the tyre, each side of the starting point, ensuring that the bead opposite the working area is always in the well. Apply lubricant as necessary. Avoid using tyre levers unless absolutely essential, to help reduce damage to the soft wheel rim. Use of the levers should be required only when the final portion of bead is to be pushed over the rim.

10   Lubricate the tyre beads again prior to inflating the tyre, and check that the wheel rim is evenly positioned in relation to the tyre beads. Inflation of the tyre may well prove impossible without the use of a high pressure air hose. The tyre will retain air completely only when the beads are pressed firmly against the rim edges at all points and it may be found when using a foot pump that air escapes at the same rate as it is pumped in. This problem may also be encountered when using an air hose on new tyres which have been compressed in storage and by virtue of their profile hold the beads away from the rim edges. To overcome this difficulty, a tourniquet may be placed around the circumference of the tyre, over the central area of the tread. The compression of the tread in this area will cause the beads to be pushed outwards in the desired direction. The type of tourniquet most widely used consists of a length of hose closed at both ends, with a suitable clamp fitted to enable both ends to be connected. An ordinary tyre valve is fitted at one end of the tube so that after the hose has been secured around the tyre it may be inflated, giving a constricting effect. Another possible method of seating beads to obtain initial inflation is to press the tyre into the angle between a wall and the floor. With the airline attached to the valve additional pressure is then applied to the tyre by the hand and shin, as shown in the accompanying illustration. The application of pressure at four points around the tyre's circumference whilst simultaneously applying the airline will often effect an initial seal between the tyre beads and wheel rim, thus allowing inflation to occur.

11   Having successfully accomplished inflation, increase the pressure to 40 psi and check that the tyre is evenly disposed on the wheel rim. This may be judged by checking that the thin positioning line found on each tyre wall is equidistant from the wheel rim around the total circumference of the tyre. If this is not the case, deflate the tyre, apply additional lubrication and reinflate. Minor adjustments to the tyre position may be made by bouncing the wheel on the ground.

12   Always run the tyres at the recommended pressures and never under- or over-inflate. The correct pressures are given in the Specifications at the start of this chapter. Note that if nonstandard tyres are fitted check with the tyre manufacturer or supplier for recommended pressures. Finally refit the valve dust cap.

8   Tubeless tyres: puncture repair and tyre renewal

1   If a puncture occurs, the tyre should be removed for inspection for damage before any attempt is made at remedial action. The temporary repair of a punctured tyre by inserting a plug from the outside should not be attempted. The manufacturers strongly recommend that no such repair is carried out on a motorcycle tyre. Not only does the tyre have a thin carcass, which does not give sufficient support to the plug, but the consequences of a sudden deflation are often sufficiently serious that the risk of such an occurrence should be avoided at all costs.

2   The tyre should be inspected both inside and out for damage to the carcass. Unfortunately the inner lining of the tyre – which takes the place of the inner tube – may easily obscure any damage and some experience is required in making a correct assessment of the tyre condition.

3   There are two main types of repair which are considered safe in repairing tubeless motorcycle tyres. The first consists of inserting a mushroom-headed plug into the hole from the inside of the tyre. The hole is prepared for insertion of the plug by reaming and the application of an adhesive. The second repair is

carried out by buffing the inner lining in the damaged area and applying a cold or vulcanised patch. Because both inspection and repair, if they are to be carried out safely, require experience in this type of work, it is recommended that the tyre be placed in the hands of a repairer with the necessary skills, rather than repaired in the home workshop.

4   In the event of an emergency, the only recommended 'get-you-home' repair is to fit a standard inner tube of the correct size. If this course of action is adopted, care should be taken to ensure that the cause of the puncture has been removed before the inner tube is fitted. It may be found that the valve hole in the rim is considerably larger than the diameter of the inner tube valve stem. To prevent the ingress of road dirt, and to help support the valve, a spacer should be fitted over the valve.

5   In the event of the unavailability of tubeless tyres, ordinary tubed tyres fitted with inner tubes of the correct size may be fitted. Refer to the manufacturer or a tyre fitting specialist to ensure that only a tyre and tube of equivalent type and suitability is fitted, and also to advise on the fitting of a valve nut to the rim hole.

9   Tubeless tyre valves: description and renewal

1   It will be appreciated from the preceding Sections that the adoption of tubeless tyres has made it necessary to modify the valve arrangement, as there is no longer an inner tube which can carry the valve core. The problem has been overcome by fitting a separate tyre valve which passes through a close-fitting hole in the rim, and which is secured by a nut and locknut. The valve is fitted from the rim well, and it follows that the valve can be removed and refitted only when the tyre has been removed from the rim. Leakage of air from around the valve body is likely to occur only if the sealing seat fails or if the nut and locknut become loose.

2   The valve core is of the same type as that used with tubed tyres, and screws into the valve body. The core can be removed with a small slotted tool which is normally incorporated in plunger type pressure gauges. Some valve dust caps incorporate a key for removing valve cores. Although tubeless tyre valves seldom give trouble, it is possible for a leak to develop if a small particle of grit lodges on the sealing face. Occasionally, an elusive slow puncture can be traced to a leaking valve core, and this should be checked before a genuine puncture is suspected.

3   The valve dust caps are a significant part of the tyre valve assembly. Not only do they prevent the ingress of road dirt in the valve, but also act as a secondary seal which will reduce the risk of sudden deflation if a valve core should fail.

10   Alternator: checking the output – US XT500 G and H models

1   The alternator fitted to XT500 G and H models is similar to that fitted to SR500 models covered in the main part of this manual (see Fig.6.4). The charging and lighting coils are an integral part of the stator assembly and cannot be renewed individually. If a fault is suspected, the charging and lighting outputs can be tested as follows.

Charging circuit

2   If the performance of the charging circuit is suspect, it should be checked using two successive tests. Check first that the battery is fully charged.

3   Connect a dc voltmeter of 0 – 20 volt range across the battery terminals. *Do not disconnect the battery leads during this test, because this will damage the electrical components.* Start the engine and allow it to run at 4000 rpm. At this speed the

voltage reading should be approximately 7 – 8 volts. Stop the engine and disconnect the test equipment. If the alternator output fails to reach this figure, the resistance of the charging coil should be checked as described below.

4   Trace the wiring back from the alternator and disconnect it from the main wiring loom. Using a multimeter set to the ohms x 1 scale, check the resistance between the black and white wire terminals on the alternator side of the block connector. Compare the reading obtained with that given in the Specifications.

5   If the reading differs widely from that specified, the charging coil must be renewed. Check first, however, that the fault is not due to a broken wire or connection; pinched or broken wires can usually be repaired by the average private owner. If the charging coil is confirmed faulty, the rotor must be removed as described in Chapter 1, Section 9 and the alternator stator assembly renewed.

6   If the charging coil resistance was found to be correct, but the dc circuit voltage was found to be unsatisfactory, the rectifier is at fault. Unfortunately specific test figures for this unit are not available. It will therefore be necessary to seek the help of a Yamaha dealer who will have the necessary equipment to test the rectifier.

Lighting circuit

7   Connect an ac voltmeter of 0 – 20 volts into the circuit. The positive lead should be connected to the yellow wire at the alternator block connector and the negative lead to the black wire. Do not separate the connector for this test.

8   Start the engine, turn on the lights and increase the engine speed to around 2000 rpm. A reading of approximately 6 – 7 volts should be obtained. Slowly increase and decrease the engine speed whilst noting the effect it has on the voltage reading. If the lighting system is in good condition the voltage reading should remain constant and should not vary greatly with engine speed. Stop the engine and disconnect the test equipment. If the output reading is found to be incorrect, or varies greatly with engine speed, the lighting coil resistance should be tested as follows.

9   Disconnect the wires from the alternator block connector and using a multimeter set to the ohms x 1 scale measure the resistance between the yellow and black wires on the alternator side of the block connector. Compare it with the reading given in the Specifications. If the reading obtained differs greatly from that specified, the lighting coil is faulty and must be renewed. Check first, however, that the fault is not due to a broken wire; pinched or broken wires can usually be repaired by the average private owner. If the coil is confirmed faulty, the rotor must be removed as described in Chapter 1, Section 9 and the alternator stator assembly renewed.

10   If the lighting coil resistance is found to be correct, but the lighting circuit voltage is found to be excessive, the regulator should be checked. Unfortunately specific test figures for the regulator are not available, and it will therefore be necessary to take the unit to a Yamaha dealer for testing or to substitute a known good unit.

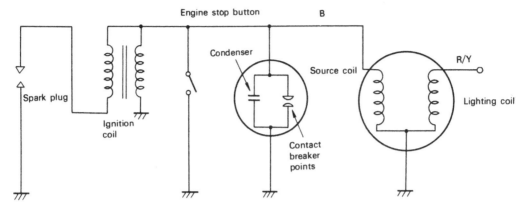

Wire color abbreviations
B: Black
R/Y: Red/Yellow

Wiring diagram – TT500 C, D and E models

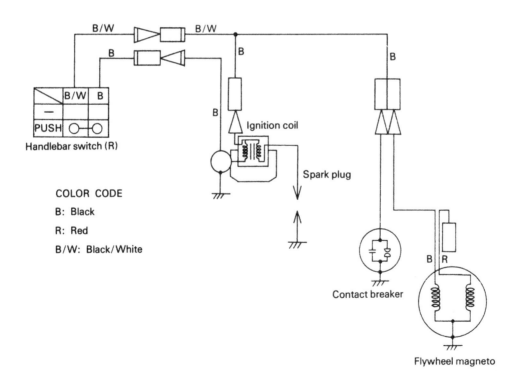

COLOR CODE
B: Black
R: Red
B/W: Black/White

Wiring diagram – TT500 F, G and H models

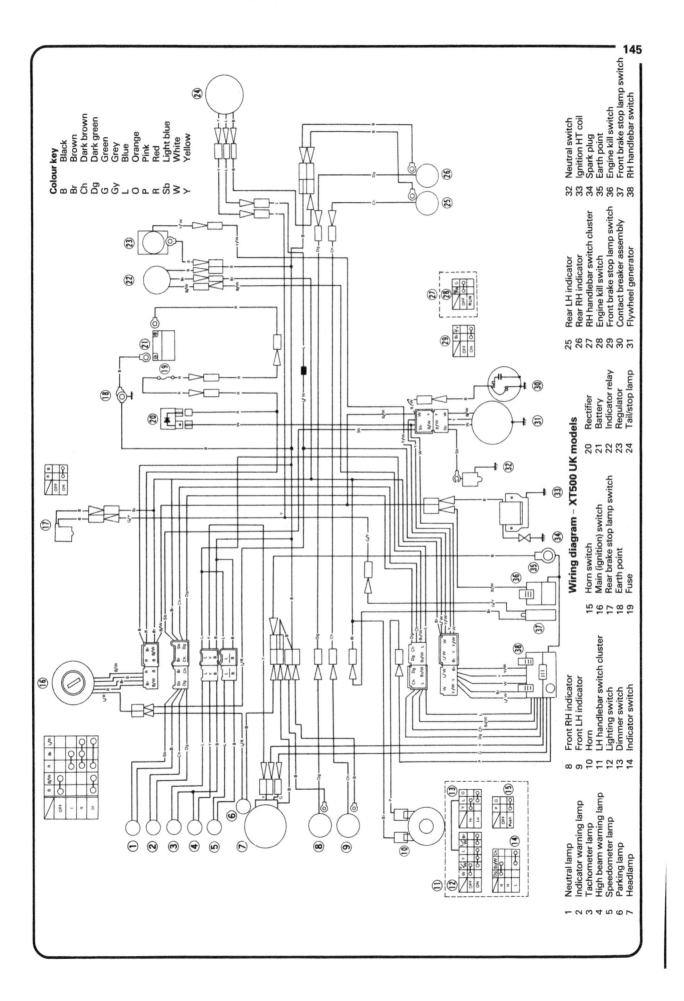

Wiring diagram – XT500 UK models

Colour key

B	Black
Br	Brown
Ch	Dark brown
Dg	Dark green
G	Green
Gy	Grey
L	Blue
O	Orange
P	Pink
R	Red
Sb	Light blue
W	White
Y	Yellow

1 Neutral lamp
2 Indicator warning lamp
3 Tachometer lamp
4 High beam warning lamp
5 Speedometer lamp
6 Parking lamp
7 Headlamp
8 Front RH indicator
9 Front LH indicator
10 Horn
11 LH handlebar switch cluster
12 Lighting switch
13 Dimmer switch
14 Indicator switch
15 Horn switch
16 Main (ignition) switch
17 Rear brake stop lamp switch
18 Earth point
19 Fuse
20 Rectifier
21 Battery
22 Indicator relay
23 Regulator
24 Tail/stop lamp
25 Rear LH indicator
26 Rear RH indicator
27 RH handlebar switch cluster
28 Engine kill switch
29 Front brake stop lamp switch
30 Contact breaker assembly
31 Flywheel generator
32 Neutral switch
33 Ignition HT coil
34 Spark plug
35 Earth point
36 Engine kill switch
37 Front brake stop lamp switch
38 RH handlebar switch

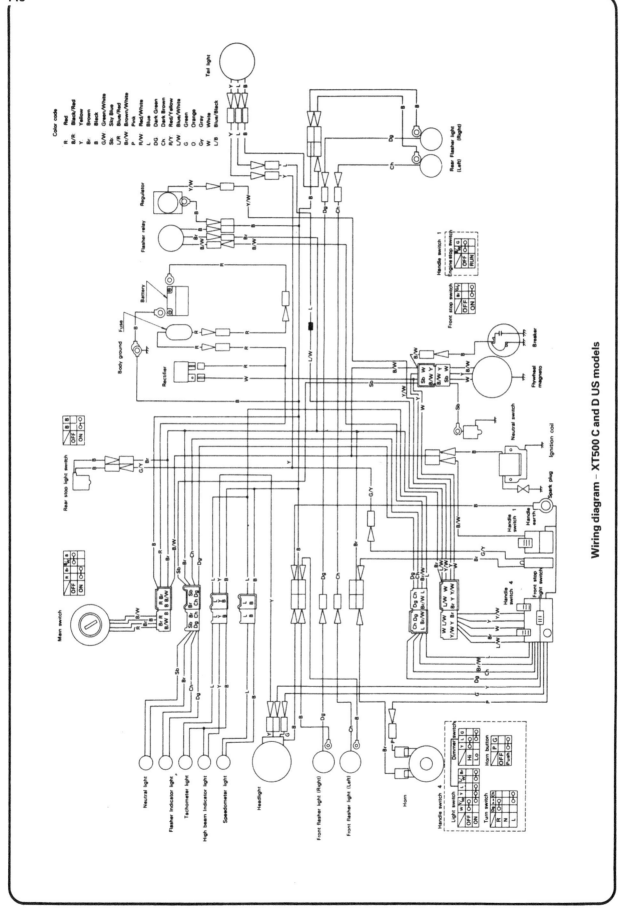

Wiring diagram – XT500 C and D US models

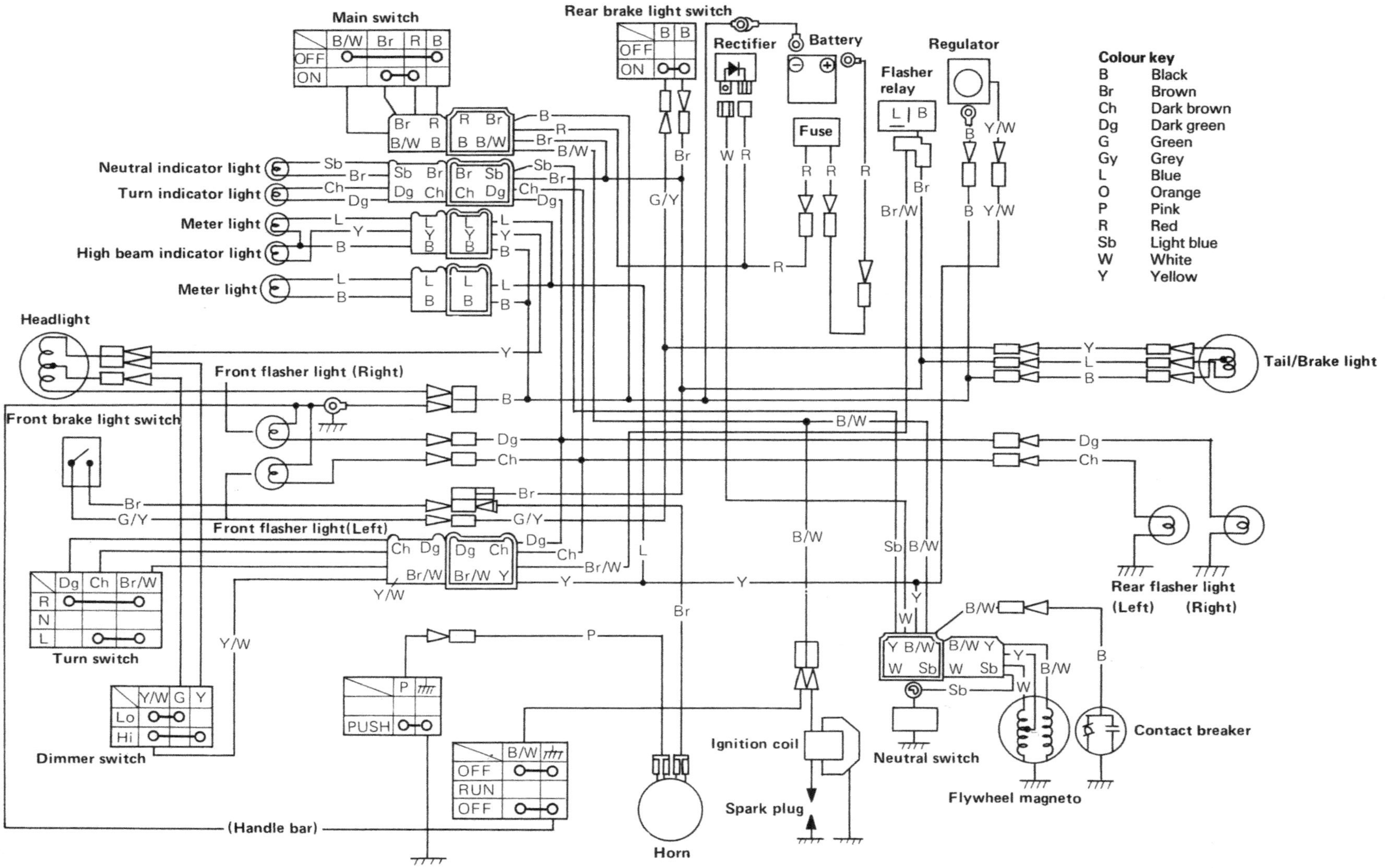

Wiring diagram – XT500 E and F US models

Colour key
B Black
Br Brown
Ch Dark brown
Dg Dark green
G Green
Gy Grey
L Blue
O Orange
P Pink
R Red
Sb Light blue
W White
Y Yellow

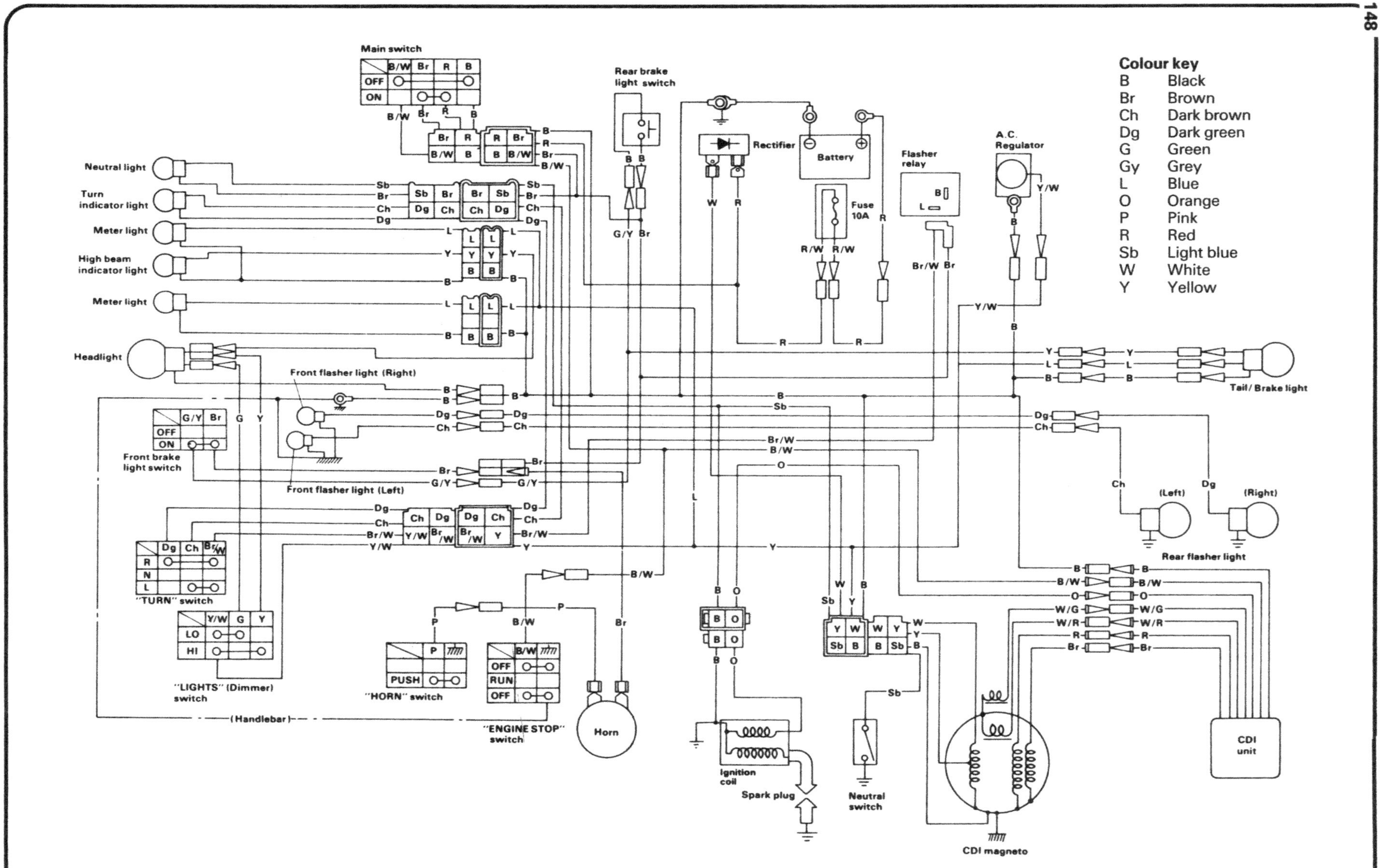

Colour key

B	Black
Br	Brown
Ch	Dark brown
Dg	Dark green
G	Green
Gy	Grey
L	Blue
O	Orange
P	Pink
R	Red
Sb	Light blue
W	White
Y	Yellow

CDI unit

Tail / Brake light

(Right)

(Left)

Rear flasher light

Rear flasher light

A.C. Regulator

CDI magneto

Flasher relay

Neutral switch

Battery

Fuse 10A

Rectifier

Ignition coil

Spark plug

Horn

Rear brake light switch

"ENGINE STOP" switch

OFF	
RUN	

"HORN" switch

PUSH	

Main switch

	B/W	Br	R	B
OFF				
ON				

Front flasher light (Right)

Front flasher light (Left)

Front brake light switch

OFF	
ON	

"TURN" switch

R		
N		
L		

"LIGHTS" (Dimmer) switch

	Y/W	
LO		
HI		

(Handlebar)

Neutral light

Turn indicator light

Meter light

High beam indicator light

Meter light

Headlight

Wiring diagram – XT500 G and H US models

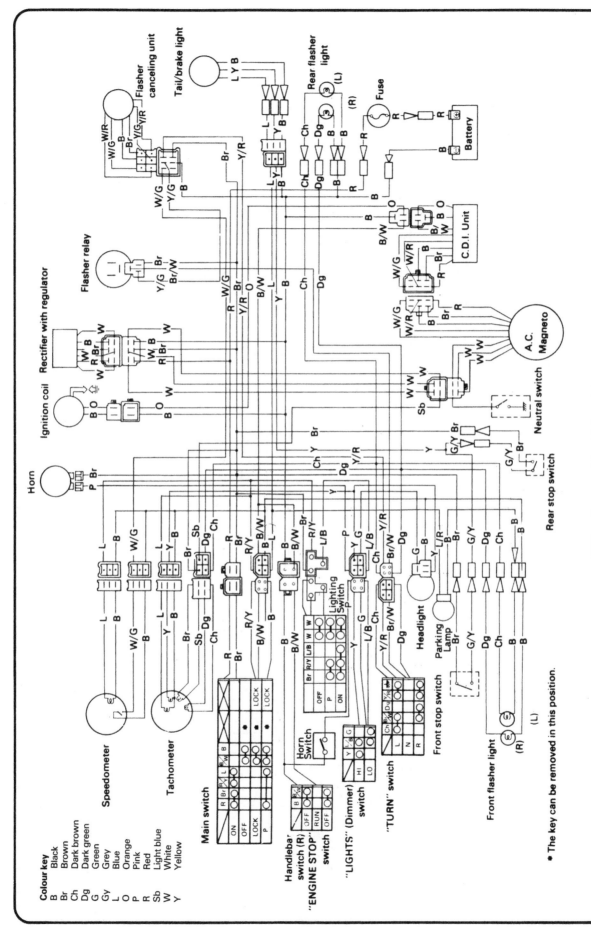

Wiring diagram – SR500 UK models

Colour key

B	Black
Br	Brown
Ch	Dark brown
Dg	Dark green
G	Green
Gy	Grey
L	Blue
O	Orange
P	Pink
R	Red
Sb	Light blue
W	White
Y	Yellow

* The key can be removed in this position.

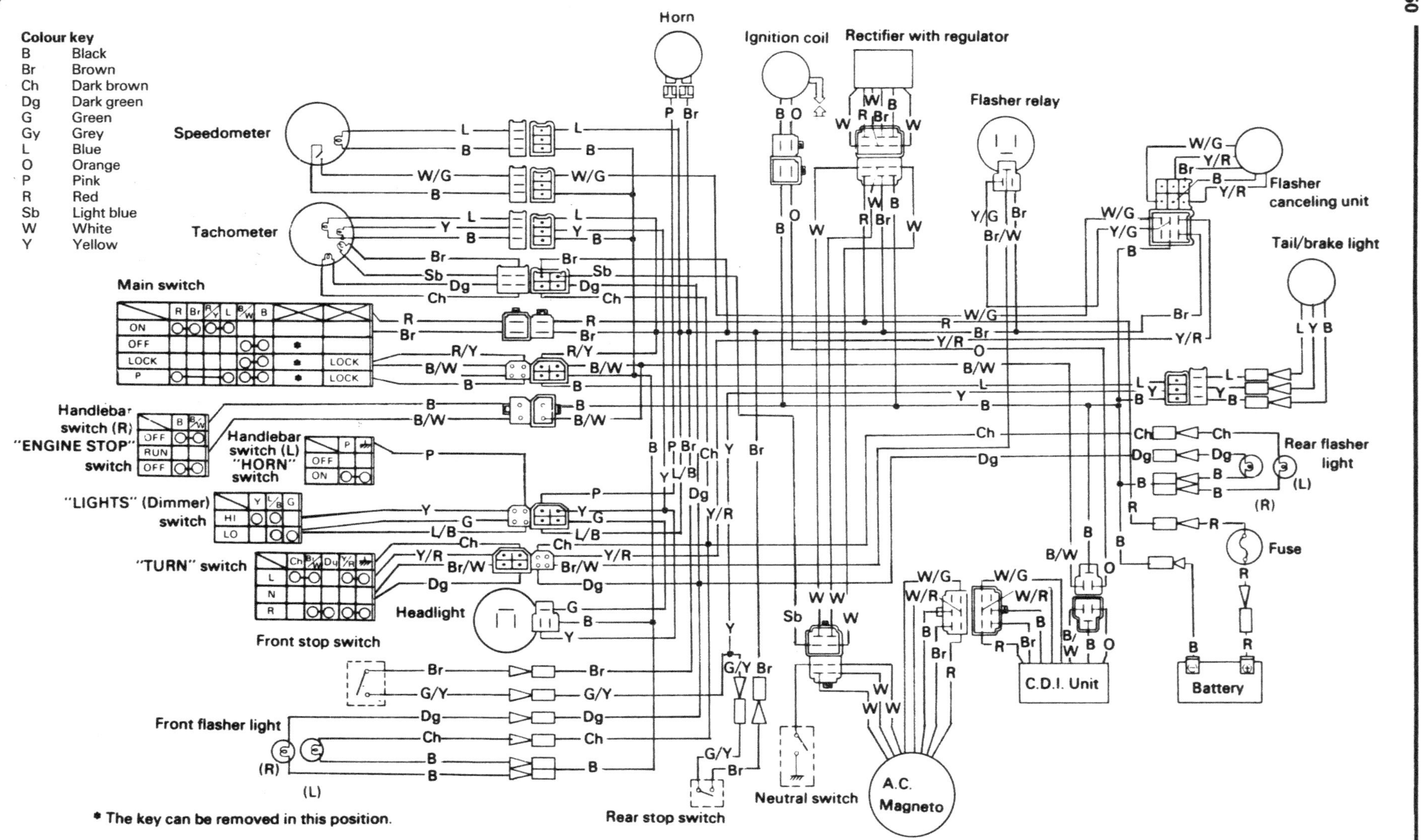

Wiring diagram – SR500 US models

Colour key
B   Black
Br   Brown
Ch   Dark brown
Dg   Dark green
G   Green
Gy   Grey
L   Blue
O   Orange
P   Pink
R   Red
Sb   Light blue
W   White
Y   Yellow

* The key can be removed in this position.

Conversion factors

Length (distance)
Inches (in)	X	25.4	=	Millimetres (mm)	X	0.0394	= Inches (in)
Feet (ft)	X	0.305	=	Metres (m)	X	3.281	= Feet (ft)
Miles	X	1.609	=	Kilometres (km)	X	0.621	= Miles

Volume (capacity)
Cubic inches (cu in; in³)	X	16.387	=	Cubic centimetres (cc; cm³)	X	0.061	= Cubic inches (cu in; in³)
Imperial pints (Imp pt)	X	0.568	=	Litres (l)	X	1.76	= Imperial pints (Imp pt)
Imperial quarts (Imp qt)	X	1.137	=	Litres (l)	X	0.88	= Imperial quarts (Imp qt)
Imperial quarts (Imp qt)	X	1.201	=	US quarts (US qt)	X	0.833	= Imperial quarts (Imp qt)
US quarts (US qt)	X	0.946	=	Litres (l)	X	1.057	= US quarts (US qt)
Imperial gallons (Imp gal)	X	4.546	=	Litres (l)	X	0.22	= Imperial gallons (Imp gal)
Imperial gallons (Imp gal)	X	1.201	=	US gallons (US gal)	X	0.833	= Imperial gallons (Imp gal)
US gallons (US gal)	X	3.785	=	Litres (l)	X	0.264	= US gallons (US gal)

Mass (weight)
Ounces (oz)	X	28.35	=	Grams (g)	X	0.035	= Ounces (oz)
Pounds (lb)	X	0.454	=	Kilograms (kg)	X	2.205	= Pounds (lb)

Force
Ounces-force (ozf; oz)	X	0.278	=	Newtons (N)	X	3.6	= Ounces-force (ozf; oz)
Pounds-force (lbf; lb)	X	4.448	=	Newtons (N)	X	0.225	= Pounds-force (lbf; lb)
Newtons (N)	X	0.1	=	Kilograms-force (kgf; kg)	X	9.81	= Newtons (N)

Pressure
Pounds-force per square inch (psi; lbf/in²; lb/in²)	X	0.070	=	Kilograms-force per square centimetre (kgf/cm²; kg/cm²)	X	14.223	= Pounds-force per square inch (psi; lbf/in²; lb/in²)
Pounds-force per square inch (psi; lbf/in²; lb/in²)	X	0.068	=	Atmospheres (atm)	X	14.696	= Pounds-force per square inch (psi; lbf/in²; lb/in²)
Pounds-force per square inch (psi; lbf/in²; lb/in²)	X	0.069	=	Bars	X	14.5	= Pounds-force per square inch (psi; lbf/in²; lb/in²)
Pounds-force per square inch (psi; lbf/in²; lb/in²)	X	6.895	=	Kilopascals (kPa)	X	0.145	= Pounds-force per square inch (psi; lbf/in²; lb/in²)
Kilopascals (kPa)	X	0.01	=	Kilograms-force per square centimetre (kgf/cm²; kg/cm²)	X	98.1	= Kilopascals (kPa)
Millibar (mbar)	X	100	=	Pascals (Pa)	X	0.01	= Millibar (mbar)
Millibar (mbar)	X	0.0145	=	Pounds-force per square inch (psi; lbf/in²; lb/in²)	X	68.947	= Millibar (mbar)
Millibar (mbar)	X	0.75	=	Millimetres of mercury (mmHg)	X	1.333	= Millibar (mbar)
Millibar (mbar)	X	0.401	=	Inches of water (inH₂O)	X	2.491	= Millibar (mbar)
Millimetres of mercury (mmHg)	X	0.535	=	Inches of water (inH₂O)	X	1.868	= Millimetres of mercury (mmHg)
Inches of water (inH₂O)	X	0.036	=	Pounds-force per square inch (psi; lbf/in²; lb/in²)	X	27.68	= Inches of water (inH₂O)

Torque (moment of force)
Pounds-force inches (lbf in; lb in)	X	1.152	=	Kilograms-force centimetre (kgf cm; kg cm)	X	0.868	= Pounds-force inches (lbf in; lb in)
Pounds-force inches (lbf in; lb in)	X	0.113	=	Newton metres (Nm)	X	8.85	= Pounds-force inches (lbf in; lb in)
Pounds-force inches (lbf in; lb in)	X	0.083	=	Pounds-force feet (lbf ft; lb ft)	X	12	= Pounds-force inches (lbf in; lb in)
Pounds-force feet (lbf ft; lb ft)	X	0.138	=	Kilograms-force metres (kgf m; kg m)	X	7.233	= Pounds-force feet (lbf ft; lb ft)
Pounds-force feet (lbf ft; lb ft)	X	1.356	=	Newton metres (Nm)	X	0.738	= Pounds-force feet (lbf ft; lb ft)
Newton metres (Nm)	X	0.102	=	Kilograms-force metres (kgf m; kg m)	X	9.804	= Newton metres (Nm)

Power
Horsepower (hp)	X	745.7	=	Watts (W)	X	0.0013	= Horsepower (hp)

Velocity (speed)
Miles per hour (miles/hr; mph)	X	1.609	=	Kilometres per hour (km/hr; kph)	X	0.621	= Miles per hour (miles/hr; mph)

Fuel consumption*
Miles per gallon, Imperial (mpg)	X	0.354	=	Kilometres per litre (km/l)	X	2.825	= Miles per gallon, Imperial (mpg)
Miles per gallon, US (mpg)	X	0.425	=	Kilometres per litre (km/l)	X	2.352	= Miles per gallon, US (mpg)

Temperature
Degrees Fahrenheit = (°C x 1.8) + 32   Degrees Celsius (Degrees Centigrade; °C) = (°F - 32) x 0.56

*It is common practice to convert from miles per gallon (mpg) to litres/100 kilometres (l/100km),
where mpg (Imperial) x l/100 km = 282 and mpg (US) x l/100 km = 235

English/American terminology

Because this book has been written in England, British English component names, phrases and spellings have been used throughout. American English usage is quite often different and whereas normally no confusion should occur, a list of equivalent terminology is given below.

English	American	English	American
Air filter	Air cleaner	Number plate	License plate
Alignment (headlamp)	Aim	Output or layshaft	Countershaft
Allen screw/key	Socket screw/wrench	Panniers	Side cases
Anticlockwise	Counterclockwise	Paraffin	Kerosene
Bottom/top gear	Low/high gear	Petrol	Gasoline
Bottom/top yoke	Bottom/top triple clamp	Petrol/fuel tank	Gas tank
Bush	Bushing	Pinking	Pinging
Carburettor	Carburetor	Rear suspension unit	Rear shock absorber
Catch	Latch	Rocker cover	Valve cover
Circlip	Snap ring	Selector	Shifter
Clutch drum	Clutch housing	Self-locking pliers	Vise-grips
Dip switch	Dimmer switch	Side or parking lamp	Parking or auxiliary light
Disulphide	Disulfide	Side or prop stand	Kick stand
Dynamo	DC generator	Silencer	Muffler
Earth	Ground	Spanner	Wrench
End float	End play	Split pin	Cotter pin
Engineer's blue	Machinist's dye	Stanchion	Tube
Exhaust pipe	Header	Sulphuric	Sulfuric
Fault diagnosis	Trouble shooting	Sump	Oil pan
Float chamber	Float bowl	Swinging arm	Swingarm
Footrest	Footpeg	Tab washer	Lock washer
Fuel/petrol tap	Petcock	Top box	Trunk
Gaiter	Boot	Torch	Flashlight
Gearbox	Transmission	Two/four stroke	Two/four cycle
Gearchange	Shift	Tyre	Tire
Gudgeon pin	Wrist/piston pin	Valve collar	Valve retainer
Indicator	Turn signal	Valve collets	Valve cotters
Inlet	Intake	Vice	Vise
Input shaft or mainshaft	Mainshaft	Wheel spindle	Axle
Kickstart	Kickstarter	White spirit	Stoddard solvent
Lower leg	Slider	Windscreen	Windshield
Mudguard	Fender		

Index